Key Enabling Technologies for Future Wireless, Wired, Optical and Satcom Applications

RIVER PUBLISHERS SERIES IN COMMUNICATIONS AND NETWORKING

Series Editors

ABBAS JAMALIPOUR
The University of Sydney
Australia

MARINA RUGGIERI
University of Rome Tor Vergata
Italy

The "River Publishers Series in Communications and Networking" is a series of comprehensive academic and professional books which focus on communication and network systems. Topics range from the theory and use of systems involving all terminals, computers, and information processors to wired and wireless networks and network layouts, protocols, architectures, and implementations. Also covered are developments stemming from new market demands in systems, products, and technologies such as personal communications services, multimedia systems, enterprise networks, and optical communications.

The series includes research monographs, edited volumes, handbooks and textbooks, providing professionals, researchers, educators, and advanced students in the field with an invaluable insight into the latest research and developments.

Topics included in this series include:-

- Communication theory
- Multimedia systems
- Network architecture
- Optical communications
- Personal communication services
- Telecoms networks
- Wi-Fi network protocols

For a list of other books in this series, visit www.riverpublishers.com

Key Enabling Technologies for Future Wireless, Wired, Optical and Satcom Applications

Editors

Björn Debaillie
imec, Belgium

Philippe Ferrari
Université Grenoble Alpes, France

Didier Belot
STMicroelectronics, France

François Brunier
Soitec, France

Christophe Gaquiere
MC2 Technologies, France

Pierre Busson
STMicroelectronics, France

Urtė Steikūnienė
Teraglobus, Lithuania

River Publishers

Routledge
Taylor & Francis Group
NEW YORK AND LONDON

Published 2024 by River Publishers
River Publishers
Alsbjergvej 10, 9260 Gistrup, Denmark
www.riverpublishers.com

Distributed exclusively by Routledge
605 Third Avenue, New York, NY 10017, USA
4 Park Square, Milton Park, Abingdon, Oxon OX14 4RN

Key Enabling Technologies for Future Wireless, Wired, Optical and Satcom Applications / Björn Debaillie, Philippe Ferrari, Didier Belot, François Brunier, Christophe Gaquiere, Pierre Busson, and Urtė Steikūnienė

Routledge is an imprint of the Taylor & Francis Group, an informa business

ISBN 978-87-7004-665-7 (hardback)
ISBN 978-87-7004-667-1 (online)
ISBN 978-8-770-04666-4 (ebook master)

Dedication

"The combined results of several people working together is often more effective than an individual scientist working alone."

John Bardeen

"Imagination is more important than knowledge. For knowledge is limited to all we now know and understand, whereas imagination embraces the entire world."

Albert Einstein

"Thoroughly conscious ignorance is the prelude to every real advance in science."

James Clerk Maxwell

Acknowledgement

The editors would like to thank all the contributors for their support in the planning and preparation of this book. The recommendations and opinions expressed in the book are those of the editors, authors, and contributors and do not necessarily represent those of any organizations, employers, or companies.

Björn Debaillie
Philippe Ferrari
Didier Belot
François Brunier
Christophe Gaquiere
Pierre Busson
Urtė Steikūnienė

Contents

Preface

Key enabling technologies for future wireless, wired, optical and satcom applications

As our interconnected world continues to expand, the importance of global innovation in communication systems and technologies grows significantly. Our increasing reliance on digital communication demands systems offering higher data traffic, providing faster and more reliable connectivity, and sustainably support a diverse range of applications. The demand for improved communication technologies has never been more urgent, necessitating substantial advancements and strategic shifts in today's development and implementation approaches.

Realizing these goals requires a transition towards higher frequency bands, such as mm-wave and sub-THz, and the adoption of disruptive technologies. These advanced frequency bands facilitate greater data transfer rates and lower latency, essential for enabling next generation communication networks. Venturing into higher frequency bands above 100 GHz for 6G further necessitates the utilisation of disruptive semiconductor technologies, including heterogeneous integration of (Bi)CMOS, SOI, and III/V components such as GaN or InP, alongside advanced packaging techniques. These innovations will play a pivotal role in achieving the objectives of ubiquitous, compact, efficient, and high-performance systems.

However, as we push the boundaries of technological possibilities, it is imperative to consider the environmental impact of these innovations. Ensuring that future communication systems are not only technologically advanced but also sustainable and responsible is paramount. This entails careful consideration of the materials used/handled, manufacturing processes employed, operational efficiency, and their recyclability. Addressing these environmental factors is essential for developing technologies that meet our communication needs while safeguarding the planet for future generations.

This book illustrates the latest research roadmaps and achievements from the European ecosystem, comprising industry, research, and academia. It focuses on the development of future wireless, wired, optical, and satellite communication (satcom) applications utilizing the mm-wave and sub-THz bands. Encompassing the entire value chain, the book explores technologies, devices, characterization, architectures, circuits, 3D heterogeneous integration, and advanced packaging techniques.

Leveraging the synergetic interactions between European CHIPS JU projects SHIFT and Move2THz, the European 3D heterogeneous integration and packaging community, and the MTT-TC9 society, this book highlights transformative developments in communication technology. The SHIFT project pioneers innovative semiconductor and packaging technologies for telecommunication areas such as 5GNR (Beyond 5G) and 6G wireless, satellite and optical networks and Earth observation. The Move2THz project is instrumental in transforming the InP platform and establishing a fully integrated European value chain, providing commercially attractive, eco-friendly, mass-market technologies suitable for sub-THz frequency operation and beyond. The IEEE MTT-TC9 society's mission is to monitor developments in emerging and technologically significant solid-state device technologies for microwave and/or mm-wave applications and promote the dissemination and sharing of technical knowledge in those areas.

The book comprehensively covers the topics discussed at the International Workshop on "Key Enabling Technologies for Future Wireless, Wired, Optical, and Satcom Applications," held at the European Microwave Week in Paris, France, on September 22, 2024. Through articles and abstracts, it provides a combined view of experts and practitioners representing academia, research, and industry in the field of wireless and optical communication systems. The workshop spans a wide range of topics, including device and substrate characterization, circuit and antenna design, and integration and packaging technologies. All of which are crucial for advancing high-performance and sustainable communication applications.

This book is a valuable resource for researchers, designers, developers, academics, post-graduate students and practitioners seeking recent research results on 5G and 6G technology. It combines several novel developments and collaborative initiatives across the entire value chain, from substrate technology up to applications. As such, it provides a comprehensive understanding of the scientific path forward and underscores the critical role of sustainable and responsible innovation in this rapidly evolving landscape.

The book is structured into 12 articles and 15 abstract descriptions, which are distributed over 5 topic clusters. Each cluster provides a comprehensive overview of the advancements and innovations within a specific domain of communication technology. Together, these clusters offer a comprehensive exploration of the latest research and developments shaping the landscape of communication technology, providing valuable insights for researchers, designers, and practitioners in the field.

The first cluster focuses on substrate technology innovations, highlighting their pivotal role in enabling emerging applications in communication systems.

- *Pascal Chevalier, Alexis Gauthier, Nicolas Guitard, Victor Milon, Frederic Monsieur, Nicolas Derrier, Claire Deglise-Favre, Didier Céli, Cédric Durand, Ophélie Foissey, Florence Sonnerat, Frederic Gianesello, Daniel Gloria*: "B55X: A SHIFT in STMicroelectronics BiCMOS Technologies" presents why BiCMOS055X is also a significant shift in STMicroelectronics BiCMOS offer, first in terms of innovation and performances, but also with respect to the device offer that has been tailored to address different applications, turning out in a flexible technology offer.
- *Yvan Morandin*: "RF advanced substrates for 5G advanced and 6G" presents the RF substrate technology roadmap, crucial for architecting the resilient, high-performance RF systems that will underpin the next gen 5G and 6G wireless infrastructure.
- *Francois Brunier*: "Advanced substrates technologies for Sub-THz era" describes how FD-SOI ecosystem innovation on III-V innovative substrates solutions can enable novel sub-THz applications.
- *Abhitosh Vais, Annie Kumar, Guillaume Boccardi, Sachin Yadav, Yves Mols, Reynald Alcotte, Bjorn Vermeersch, Mark Ingels, Uthayasankaran Peralagu, César Roda Neve, Bruno Ghyselen, Bertrand Parvais, Piet Wambacq, Bernardette Kunert, Nadine Collaert*: "A CMOS compatible III-V-on-300mm Si technology for future high speed communication systems" arguments for the upscaling of III-V technology on to 300mm Si platforms and presents how to achieve its integration into existing CMOS platform.
- **Romain Hersent*: "InP-on-Si Technologies For Next Generation Optical Communication High-Speed Analog Front-Ends" discuss the challenges related to the design of large-swing, high-efficiency and extreme-high-symbol-rate analog electronics integrated circuits for next generation >1.6T optical /6G communication systems.
- **Kimon Vivien*: "150nm Gallium Nitride on Silicon Carbide Technology for High Power 5G New Radio Applications" describes the 150nm Gallium Nitride on Silicon Carbide technology of UMS, with a focus on high power, sub-6 GHz applications.

The second cluster delves deeper into advancing high-frequency device technology modelling and enhancing the sustainability of electronics, addressing critical challenges in the field.

- **Joff Derluyn*: "Post-process substrate porosification for RF applications" proposes a process to the handle silicon substrate after and independently from the fabrication of the front side RF circuitry to achieve extremely low harmonic distortion and microwave losses, far out-performing RF-SOI.
- *Bruno Sagnes, Ayenew Adebabay Belie, Menel Bouhouche, Johnny El Beyrouthy, Fabien Pascal, Jérôme Boch, Pascal Chevalier, Daniel Gloria*: "The impact of irradiation on DC characteristics and Low Frequency Noise of advanced SiGe:C HBTs" presents their investigated X-ray and gamma total ionizing dose impacts on advanced SiGe:C HBTs supplied by STMicroelectronics.
- *Yewulsew Manale Wodaje, Bachar Baridi, Olivier Occello, François Danneville, Pascal Szriftgiser, Guillaume Ducournau*: "D-band modulated signal generation using photonics techniques" addresses a photonics approach to generate modulated signals in the range 110-170 GHz.
- *Gunther Walden*: "Decarbonizing the electronics industry to achieve Net Zero" describes essential questions and approaches to address the United Nations net-zero emissions target for 2050.

In the third cluster, innovative circuit design and architectures for high-speed communication are explored, showcasing cutting-edge developments in this rapidly evolving area.

- **Hassan Aboushady*: "A multi-standard RF Bandpass Sigma-Delta ADC" presents a 65 nm CMOS LC resonator-based tunable bandpass RF Sigma-Delta modulator with a centre frequency from 1.5 to 3.0 GHz and a corresponding sampling frequency from 6.0 to 12.0 GHz.
- *Oliver Hauck, Fred Buchali, Michael Collisi, Michael Möller*: "Analog Multiplexing for Bandwidth and Sampling Rate Multiplication of DACs in Coherent Optical Transmission Systems" presents 2:1 and 4:1 analog time division multiplexing circuit concepts, a frequency domain ENoB-model approach and measurement results.
- **Cédric Dehos*: "D-Band RF architecture for Beyond 5G wireless networks" discusses the challenges of D-band transceiver design with integrated scalable and reconfigurable antennas.

- **Leonardo Gomes*: “E-band & D-band VCOs: distributed tank design methodology, bufferless approach” proposes a new methodology to design mm-wave VCOs based on distributed resonators using a global optimization routine.

The fourth cluster examines technology-level heterogeneous integration approaches, demonstrating how these strategies can enhance characteristics and performance across various applications.

- **Olivier Valorge*: “2.5D, 3D assembly technologies for RF, mmW and sub-THz heterogeneous systems” described different 2.5D & 3D integration solutions from the simplest to the most advanced while focusing on mmW/subTHz heterogeneous integrated systems.
- **Hady Yacoub*: “Heterointegration approaches for InP-HBT technologies for 5G applications and beyond” presents an overview of heterointegration techniques, their RF performance, and their scalability roadmap towards higher frequencies.
- **Siddhartha Sinha*: “RF-Heterointegration at Wafer-level and Panel-level for mm-wave Applications” presents different mm-wave packaging technologies, including RF-optimized waferscale silicon interposer technologies and PCB embedded Air-Filled Waveguides.
- *Bertrand Ardouin, Tom Keinicke Johansen, Antoine Chauvet, Romain Hersent, Virginie Nodjiadjim, Agnieszka Konczykowska, Nil Davy, Muriel Riet, Colin Mismer*: “Challenges for 2.5D and 3D integration of InP HBT technology” reviews the key specificities of InP DHBT technologies and highlights their 2.5 and 3D integration challenges.
- **Frederic Gianesello*: “SiGe BiCMOS & III-V technologies heterogeneous Integration challenges” reviews different heterogeneous integration with III-V technologies and their associated challenges by discussing SiGe BiCMOS heterogeneous integration with GaN on SiC and InP technologies.

Finally, the fifth cluster explores disruptive system design, integration, and packaging technologies, particularly focusing on their role in enabling future (sub-)THz and optical applications.

- **Tanja Braun*: “Advanced Packaging Solutions for mmWave Applications” introduces advanced packaging solutions suitable for RF and mmWave applications and antenna-in-package solutions, shielding on package level, passive component integration or heterogeneous integration of III/V semiconductors with Si or SiGe in one package.

- **Mikko Varonen*: "Modular 3D mmW and THz packaging concepts and technologies" focusses on packaging techniques for efficient phased array scaling and VTT's 3D modular packaging concepts for heterogeneous integration of different MMIC technologies.
- *Hiroaki Takahashi, Pelin Suealp, Erich Schlaffer*: "Analysis of Quasi-Coaxial Via implemented in IC Substrate using Multiple Scattering Method" presents numerical simulation- and measurement-based characterization of interconnections implemented in organic IC-substrates for 6G communication.
- **Abdel Hadi Hobballah*: "LDS and AMP processes for RF antenna in package AiP applications in the E and D-band" discusses the LDS laser direct structuring technology for the manufacturing of advanced component packaging for Antenna-in-Package (AiP) solutions operating within the E and D bands.
- **Akanksha Bhutani*: "Sub-THz Antenna and Package Integration for Miniaturized Surface Mount Device Modules" describes antenna and package developments focussing on miniaturized surface-mount devices in the WR6 (110 – 170 GHz), WR3 (220 – 325 GHz) and WR2.2 (325 – 500 GHz) bands.
- *Alessandro Fonte, Stefano Moscato, Riccardo Moro, Andrea Pallotta, Andrea Mazzanti, Andrea Bilato, Guglielmo De Filippi, Lorenzo Piotto, Francesco Centurelli, Pietro Monsurrò, Hassan Sadeghi Chameh, Pasquale Tommasino, Alessandro Trifiletti, Daniele Lodi Rizzini, Francesco Tesolin, Simone Mattia Dartizio, Salvatore Levantino*: "D-band Phased Array Antenna Module for 5G Backhaul" presents advancements of the basic element of the D-band phased array antenna system through a mockup of the module and the package, the D-band antenna concept and the RFICs developed by using the cutting-edge STMicroelectronics 55nm and 55X SiGe BiCMOS technologies.
- *Giovanni Mangraviti, Cédric Dehos, Vincent Puyal, Olivier Richard, Ghita Yaakoubi khbiza, Francesco Foglia Manzillo, Xuan Viet Linh Nguyen, Francesco Filice, Ettore Noccetti, Pierre-Louis Hellier, Florence Podevin, Sylvain Bourdel, Andrea Ruffino, Kyung-Sik Choi , Basem Abdelaziz Abdelmagid, Mohamed Eleraky, Hua Wang, Bertrand Ardouin, Tom, Keinicke Johansen, Luca Fanori, Qiuting Huang, Erich Schlaffer, Piet Wambacq, Dominique Morche, Björn Debaillie*: "Sub-THz transceiver design for future generation mobile communications"

presents an innovative D-band transceiver circuit and module prototype design based on the most advanced BiCMOS and InP technologies.

- *Francesco Manni, Paolo Colantonio, Rocco Giofrè, Ernesto Limiti, Patrick Ettore Longhi, Steven Mejillones Caicedo, Stefano Moscato, Alessandro Fonte*: "Ka-Band GaN-on-SiC Power Amplifier for High EIRP Satellite Phased Antenna Array" presents an innovative a phased antenna array operating in the Ka-band, from 25.5 to 27 GHz enabling the adoption of a custom-designed power amplifier based on 0.15 μm GaN-on-SiC technology capable of delivering ~30% PAE and 28 dBm output power.

* The abstracts are compiled in the appendix of this book.

Editors' Biography

1st Editor: Björn Debaillie, imec, Belgium

Björn Debaillie (Senior, IEEE) leads imec's collaborative R&D program on cutting-edge connected computing, covering high speed communications, high resolution sensing, and neuromorphic computing. As a seasoned researcher and manager, he is responsible for strategic collaborations and partnerships, innovation management, and public funding policies as well as the operational management and coordination across imec's collaborative programs and projects. Björn Debaillie coordinates public funded projects and seeds new initiatives. He holds patents, received awards and authored books and international papers published in various journals and conference proceedings.

2nd Editor: Philippe Ferrari, Université Grenoble Alpes, France

Philippe Ferrari (Senior, IEEE) received his Ph.D. degree from the "Institut National Polytechnique de Grenoble" (INPG), France, in 1992, with honors. Since 2004, he has been a professor at the University Grenoble Alpes, Grenoble, France. His main research interests concern tunable and miniaturized devices, such as filters, phase shifters, matching networks, couplers, power dividers and VCOs. These devices are developed in many technologies, PCB, CMOS/BiCMOS, and nanowires, at RF and mm-wave frequencies. He has worked towards the development of slow-wave CPW, and developed new topologies of slow-wave transmission lines, based on microstrip lines and SIWs, respectively. He is author or co-author of more than 300 papers published in international journals or conferences, and co-holder of six patents. He is a member of the Editorial Board of the International Journal on RF and Microwave Computer-Aided Engineering (Wiley), an Associate Editor of the

International Journal of Microwave and Wireless Technologies (EuMA) and member of the Editorial Board of Electronics Letters.

3rd Editor: Didier Belot, STMicroelectronics, France
Didier Belot first spent 30 years in ST Technology R&D, where he worked initially on characterization-modeling of bipolar transistors before moving to high frequency analog design to develop optical-electronic interfaces. In 1995, he moved to Analog-RF design to work on cellular and Bluetooth transceivers, in SiGe and CMOS respectively. In 2006, he created a joint team with LETI, to initiate the development of mmW R&D prototypes at 60 GHz in CMOS-65, which led to a demonstration of a 4 Gbs wireless link at 60 GHz over 1 m. In 2014, he joined CEA-LETI, to continue his research work, on mmW CMOS, on mmW plastic guides, and, on very high speed mmW (1–10 GBs and more). He has also played an important role in the orientation of the RF and mmW roadmaps, and in the management of programs such as IPCEI within LETI. At the beginning of 2023 he came back to ST in the Technology Design Platform group, in charge of wireless strategy and innovation. Moreover, he has participated and is participating in the elaboration of European and worldwide Roadmaps on Wireless, (NEREID, CORENECT, ECS-SRIA, IRDS, etc). His current research interests include millimeter wave propagation through plastic, sub-THz communications, III-V devices on silicon for mmW and THz applications, and the use of RF for quantum computing. He was a member of the French National Scientific Council "Micro and Nanotechnologies" from 2012 to 2016, a member of various conference program committees such as RFIC, ESSCIRC, ISSCC, IEDM. He is a member of the IEEE-MTT-9 (mmW and THz Devices to System) technology committee and the EuMW technical committee. He is a reviewer for the IEEE MTT and SSC journals, and author or co-author of more than 400 publications and 70 patents.

4rd Editor: François Brunier, SOITEC, France
François Brunier (Member, IEEE) graduated as a physics and electronics engineer from Centrale-Supelec in 1997. From 1998 to 2002, he worked as device integration engineer for embedded DRAM products in STMicroelectronics Crolles. In 2002, he joined Soitec as head of advanced characterization

laboratory. From 2009 to 2011, as a product manager, he led the RF-SOI and power SOI product development and offering. Since 2012, as a partnership program manager, he is in charge of European collaborative Chips JU programs, IPCEI and public relations.

5th Editor: Christophe Gaquiere, MC2-Technologies, France

Christophe Gaquière is currently full professor at the University of Lille and carries out his research activity at the Institut d'Electronique de Microélectronique et de Nanotechnology (IEMN). The topics concern design, fabrication, characterization and modeling of HEMT's and HBT devices. He works on GaAs, InP, metamorphic HEMT's and now he is involved in the GaN activities. His main activities are microwave characterizations (small and large signal between 1 and 500 GHz) in order to correlate the microwave performances with the technological and topology parameters. Today, his activities concern mainly the investigation of two-dimensional electronic plasmons and gunn like effects for THz solid state GaN based detectors and emitters (HEMT and SSD), AlGaN/GaN nano-wires for microwave applications and MEMS activities based also on GaN. He was responsible for the microwave characterization part of the common laboratory between Thales TRT and IEMN focus on wide band gap semiconductor (GaN, SiC, and Diamond) from 2003 up to 2007. He is in the TPC of several European conferences. He was in charge of the silicon millimeter wave advanced technologies part of the common lab between ST microelectronics and IEMN. He co-founded the company MC2-technolgies in 2004 (95 people) and he is general manager. Christophe Gaquière is the author or co-author of more than 150 publications and 300 communications.

6th Editor: Pierre Busson, STMicroelectronics, France

Pierre Busson (Senior, IEEE) received his M.Sc. degree from the "Ecole Centrale d'Electronique" of Paris France in 1985 and his Ph.D. degree from the University of Rennes, France. In 1993, he worked for CELAR, national military research center, where he was involved in realization of propagation channel simulator for spread spectrum and frequency hopping. In 1995, he worked for CNET Rennes, national telecommunication research center, for the realization of the first integrated demonstrator for terrestrial digital TV with COFDM link. In 1997, he joined STMicroelectronics where he was involved in set-top boxes circuits for terrestrial, satellite and cable. He is currently Wireless RF System/Architect and a Fellow of ST Technical Staff, working on the development of the next generation for wireless system.

7th Editor: Urtė Steikūnienė, Teraglobus, Lithuania
Urtė Steikūnienė earned her M.S.S. degree from the University of Warwick and began her career as a public sector auditor for PriceWaterhouseCoopers in London. In 2019, she co-founded a start-up focused on developing an FDA-registered medical device for diabetes patients. She holds four patents and is currently a Senior Project Manager at Teraglobus, where she oversees the execution of large collaborative projects co-funded by the Chips Joint Undertaking and Horizon Europe programmes.

List of Contributors

Abdelaziz Abdelmagid, Basem, *ETH Zürich, Switzerland*

Aboushady, Hassan, *Seamless Waves, France*

Adebabay Belie, Ayenew, *Université de Montpellier, France*

Alcotte, Reynald, *imec, Belgium*

Ardouin, Bertrand, *III-V Lab, France*

Baridi, Bachar, *CNRS - Université de Lille, France*

Bhutani, Akanksha , *Karlsruher Institut für Technologie (KIT), Germany*

Bilato, Andrea, *Università di Pavia, Italy*

Blache, Fabrice, *III-V Lab, France*

Boccardi, Guillaume, *imec, Belgium*

Boch, Jérôme, *Université de Montpellier, France*

Bouhouche, Menel, *Université de Montpellier, France*

Bourdel, Sylvain, *Grenoble INP, France; Université Grenoble Alpes, France*

Boutry, Hervé, *CEA-Leti, France*

Braun, Tanja , *Fraunhofer IZM, Germany*

Brunier, François, *Soitec, France*

Buchali, Fred, *Nokia Network Infrastructure, Germany*

Caicedo Mejillones, Steven, *SIAE MICROELETTRONICA, Italy*

Céli, Didier, *STMicroelectronics, France*

Centurelli, Francesco, *Università di Roma "La Sapienza", Italy*

Chauvet, Antoine, *III-V Lab, France*

Chevalier, Pascal, *STMicroelectronics, France*

Choi , Kyung-Sik, *ETH Zürich, Switzerland*

Colantonio, Paolo, *Università di Roma "Tor Vergata", Italy*

Collaert, Nadine, *imec, Belgium; Vrije Universiteit Brussel, Belgium*

Collisi , Michael, *Universität des Saarlandes, Germany*

Danneville, François, *CNRS - Université de Lille, France*

Dartizio, Simone Mattia , *Politecnico di Milano, Italy*

Davy, Nil, *III-V Lab, France*

De Filippi, Guglielmo, *Università di Pavia, Italy*

Debaillie, Björn, *imec, Belgium*

Deglise-Favre, Claire, *STMicroelectronics, France*

Dehos, Cédric, *CEA-Leti, France*

Derluyn, Joff, *Incize, Belgium*

Derrier, Nicolas , *STMicroelectronics, France*

Ducournau, Guillaume, *CNRS - Université de Lille, France*

Durand, Cédric, *STMicroelectronics, France*

Duval, Bernadette, *III-V Lab, France*

El Beyrouthy, Johnny , *Université de Montpellier, France*

Eleraky, Mohamed, *ETH Zürich, Switzerland*

Fanori, Luca, *Advanced Circuit Pursuit (ACP), Switzerland*

Filice, Francesco, *imec, Belgium*

Foglia Manzillo, Francesco, *CEA-Leti, France*

Foissey, Ophélie, *STMicroelectronics, France*

Fonte, Alessandro , *SIAE MICROELETTRONICA, Italy*

Gauthier, Alexis, *STMicroelectronics, France*

Ghyselen, Bruno, *Soitec, France*

Gianesello, Frederic, *STMicroelectronics, France*

Giofrè, Rocco, *Università di Roma "Tor Vergata", Italy*

Gloria, Daniel, *STMicroelectronics, France*

Goix, Michel, *III-V Lab, France*

Gomes, Leonardo, *Université Grenoble Alpes, France; ST Microelectronics, France*

Guitard, Nicolas, *STMicroelectronics, France*

Haizmann, Andreas, *CircularTree, Germany*

Hauck, Oliver , *Universität des Saarlandes, Germany*

Hellier, Pierre-Louis, *Université Grenoble Alpes, France*

Hersent, Romain, *III-V Lab, France*

Hobballah, Abdel Hadi, *S2P-Smart Plastic Products, France*

Huang, Qiuting , *Advanced Circuit Pursuit (ACP), Switzerland*

Ingels, Mark, *imec, Belgium*

Johansen, Tom Keinicke, *Danmarks Tekniske Universitet, Denmark*

Jorge, Filipe, *III-V Lab, France*

Konczykowska, Agnieszka, *III-V Lab, France*

Kumar, Annie, *imec, Belgium*

Kunert, Bernardette, *imec, Belgium*

Levantino, Salvatore, *Politecnico di Milano, Italy*

Limiti, Ernesto, *Università di Roma "Tor Vergata", Italy*

Lodi Rizzini, Daniele , *Politecnico di Milano, Italy*

Longhi, Patrick, *Università di Roma "Tor Vergata", Italy*

Mangraviti, Giovanni , *imec, Belgium*

Manni, Francesco, *Università di Roma "Tor Vergata", Italy*

Mazzanti, Andrea, *Università di Pavia, Italy*

Milon, Victor, *STMicroelectronics, France*

Mismer, Colin, *III-V Lab, France*

Möller, Michael , *Universität des Saarlandes, Germany*

Mols, Yves, *imec, Belgium*

Monsieur, Frederic, *STMicroelectronics, France*

Monsurrò, Pietro, *Università di Roma "La Sapienza", Italy*

Morandini , Yvan, *Soitec, France*

Morche, Dominique, *CEA-Leti, France*

Moro, Riccardo, *SIAE MICROELETTRONICA, Italy*

Moscato, Stefano, *SIAE MICROELETTRONICA, Italy*

Nguyen, Xuan Viet Linh, *CEA-Leti, France*

Noccetti, Ettore, *imec, Belgium; Università di Pisa, Italy*

Nodjiadjim, Virginie, *III-V Lab, France*

Occello, Olivier, *CNRS - Université de Lille, France*

Pallotta, Andrea, *STMicroelectronics, Italy*

Parvais, Bertrand, *imec, Belgium; Vrije Universiteit Brussel, Belgium*

Pascal, Fabien, *Université de Montpellier, France*

Peralagu, Uthayasankaran, *imec, Belgium*

Piotto, Lorenzo, *Università di Pavia, Italy*

Podevin, Florence , *Grenoble INP, France*

Puyal, Vincent, *CEA-Leti, France*

Richard, Olivier, *CEA-Leti, France*

Riet, Muriel, *III-V Lab, France*

Roda Neve, César, *Soitec, France*

Ruffino, Andrea, *ETH Zürich, Switzerland*

Sadeghi Chameh, Hassan, *Università di Roma "La Sapienza", Italy*

Sagnes, Bruno, *Université de Montpellier, France*

Schlaffer, Erich, *AT&S, Austria*

Sinha, Siddhartha , *imec, Belgium*

Sonnerat, Florence , *STMicroelectronics, France*

Suealp , Pelin, *AT&S, Austria*

Szriftgiser, Pascal, *CNRS - Université de Lille, France*

Takahashi, Hiroaki, *AT&S, Austria*

Tesolin, Francesco, *Politecnico di Milano, Italy*

Tommasino, Pasquale, *Università di Roma "La Sapienza", Italy*

Trifiletti, Alessandro, *Università di Roma "La Sapienza", Italy*

Vais, Abhitosh , *imec, Belgium*

Valorge, Olivier, *CEA-Leti, France*

Varonen, Mikko, *VTT Technical Research Centre of Finland, Finland*

Vermeersch, Bjorn, *imec, Belgium*

Vivien, Kimon, *United Monolithic Semiconductors (UMS), France*

Walden, Gunther, *CircularTree, Germany*

Wambacq, Piet, *imec, Belgium; Vrije Universiteit Brussel, Belgium*

Wang, Hua, *ETH Zürich, Switzerland*

Wodaje, Yewulsew Manale, *CNRS - Université de Lille, France*

Yaakoubi khbiza, Ghita, *CEA-Leti, France*

Yacoub, Hady, *Ferdinand-Braun-Institut, Germany*

List of Figures

Chapter 2

Chapter 3

List of Tables

B55X: A SHIFT in STMicroelectronics BiCMOS Technologies

P. Chevalier, *Member, IEEE*, A. Gauthier, N. Guitard, V. Milon, F. Monsieur, N. Derrier, C. Deglise-Favre, D. Céli, C. Durand, O. Foissey, F. Sonnerat, F. Gianesello, D. Gloria

Abstract—**The SHIFT (Sustainable tecHnologies enablIng Future Telecom applications) project funded by Chips JU and National Authorities partly supports the qualification towards the production of STMicroelectronics BiCMOS055X, which is one of the key technologies evaluated in this project. This paper examines why BiCMOS055X is also a significant shift in STMicroelectronics BiCMOS offer, first in terms of innovation and performances, but also with respect to the device offer that has been tailored to address different applications, turning out in a versatile technology offer.**

Index Terms—**BiCMOS, silicon germanium, HBT, RF, THz, communication, optical, wireless, Satcom.**

I. INTRODUCTION

THE first consumer market of silicon germanium (SiGe) BiCMOS technology was the radiofrequency (RF) transceiver of cellular phones about 20 years ago. This application, as the TV tuner one, has been progressively captured by complementary metal–oxide–semiconductor (CMOS) technology [1], which is currently happening for automotive radars. On the other hand, whereas the RF performance of CMOS does not improve anymore, applications served today by SiGe BiCMOS, i.e., RF front-end modules (FEM) and optical transceivers, benefit from the intrinsic superiority of the SiGe heterojunction bipolar transistor (HBT) [2]. In addition, there are perspectives to improve performances of SiGe HBT beyond what is currently being qualified for production [3][4]. However, although the RF performance of the overall platform, i.e., including all the devices, is of paramount importance, technological complexity and related cost must not be neglected. It is increasingly challenging to serve multiple applications featuring different cost/performance trade-offs. Indeed, low earth orbit (LEO) satellite communications (Satcom) in Ku and Ka bands, i.e., between 10 and 40 GHz, which generate consumer-like production volumes [5], put severe constraints on wafer cost, while being quite demanding on performance. Optical transceivers, whose demand is expected to increase with the emergence of artificial intelligence (AI) clusters [6], and targeted data rates for next generations are 200 and 400 G per lane, require additional options. Finally,

This work received funding from Chips Joint Undertaking (Chips JU) under grant agreement n° 101096256 (project SHIFT). The Chips JU receives support from the European Union's Horizon Europe research and innovation program and the National Authorities.

It has been partially financed by the French Public Authorities within France 2030 and in the frame of the IPCEI ME/CT.

All authors are with STMicroelectronics, 850 rue Jean Monnet, 38926 Crolles, France (e-mail: pascal.chevalier@st.com).

RF FEMs for future 5G+ and 6G wireless communication infrastructures push the technology to its operation frequency limits, which is today the D-band, i.e., between 140 and 170 GHz. Practically, the core of the technology, and more especially the SiGe HBT architecture, is not modified to address the different applications although the transistor is used at very different bias conditions. But the technology content must be adapted to get the best well-known power–performance–area–cost (PPAC) key performance indicator (KPI). The "S" of "sustainability," a priority for STMicroelectronics (ST) [7], must be added. It is also part of the SHIFT (Sustainable tecHnologies enablIng Future Telecom applications) project [8], which relies on BiCMOS055X (B55X) technology. B55X, which serves the Move2THz [9] project too, is the topic of this paper. The first part reviews the history of silicon germanium BiCMOS technology development at ST. It allows understanding, in the second part, why B55X is a disruptive technology in the ST roadmap with respect to innovation, performance, and versatility. The third part discusses the positioning of this technology versus the current state-of-the-art. Finally, the key points of the technology are summarized in the conclusion and perspectives for next generations are drawn.

II. HIGH-SPEED SIGE BICMOS TECHNOLOGIES HISTORY AT STMICROELECTRONICS

ST holds 25+ years of experience in SiGe BiCMOS illustrated in Figure 1, showing the increase of SiGe HBT speed over the generations and how it has been combined with the CMOS nodes to serve different markets. It started with a 0.35-µm BiCMOS technology (BiCMOS6G) [10] and continued with two generations in 0.25 µm (BiCMOS7 [11] and BiCMOS7RF [12]) and two generations in 0.13 µm (BiCMOS9 [13] and BiCMOS9MW [14]). These technologies were developed on 200-mm wafers. The sixth and seventh generations are based on 55-nm CMOS and thus migrated to 300-mm wafers, first with BiCMOS055 [15] and now with BiCMOS055X [16]. Table 1 summarizes the main process differences between the generations.

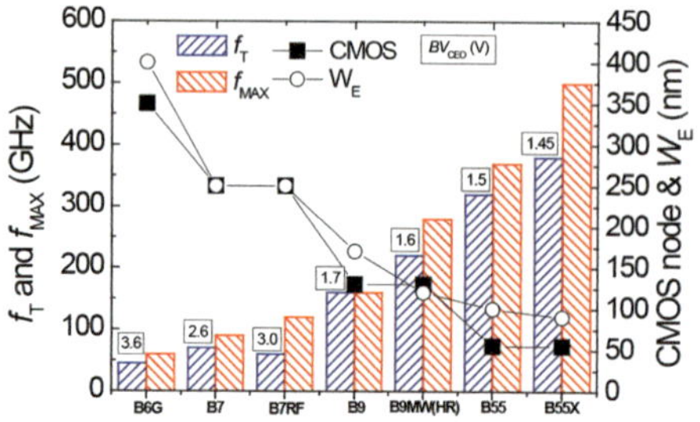

Fig. 1. STMicroelectronics high-speed SiGe BiCMOS technologies (HBT performance & CMOS node). W_E is the physical emitter width. BV_{CEO} is the emitter–collector breakdown voltage of the HBT with an open base. f_T and f_{MAX} correspond to the peak values measured for the SiGe HBT.

Beyond the CMOS node, which did not always change between two generations, the SiGe HBT architecture evolved continuously. It started from a quasi-self-aligned (QSA) single-polysilicon (SP) architecture using a non-selective epitaxy growth (NSEG) of the SiGe base [10]. This architecture was simple but suffered from several limitations, the main one being related to the implantation of the extrinsic base. The second generation of SiGe HBT solved this issue with the introduction of a double-polysilicon (DP) architecture [11]. In addition, the introduction of carbon in the base [12] and deep trenches isolation (DTI) helped to further improve the performance. The full self-alignment (FSA) between the emitter and the base, a major step in base resistance (R_B) reduction [17], was brought by the third generation [14][15].

TABLE I
DIFFERENCES BETWEEN STMICROELECTRONICS BICMOS TECHNOLOGIES

Technology		B6G	B7	B7RF	B9	B9MW		B55	B55X
CMOS node (nm)		350	250	250	130	130		55	55
Bulk substrate resistivty		SR	SR	SR	SR	SR	HR	SR	SR
SiGe HBT architecture	**Emitter-Base**	QSA-SP		QSA-DP		FSA-DP			EXBIC
	Intrinsic base epitaxy	NSEG SiGe		NSEG SiGe:C		SEG SiGe:C			
	Extrinsic base epitaxy	No							Yes
	Collector epitaxy	NSEG							SEG
	DTI	No	Yes						Optional
BEOL		Digital (native)		Thick Cu	Digital (native)	Optimized for RF (Thick Cu native)			

It was achieved by moving to the selective epitaxial growth (SEG) of the base. The fourth generation addressed the critical point of the intrinsic-to-extrinsic base link resistance using a specific epitaxy process step. It is achieved with the EXBIC (*Epitaxial eXtrinsic Base Isolated from the Collector*) architecture developed for the B55X technology [18]. As illustrated in Figure 1, where emitter–collector breakdown voltage (BV_{CEO}) evolution is shown, peak frequency increases have been achieved at the expense of the breakdown voltage (collector doping is increased to delay the onset of the Kirk effect). However, it decreased much less than peak frequencies increased. In addition, the base–collector breakdown voltage (BV_{CBO}), the maximum voltage at which the transistor can be biased, is about 3–4 times higher than BV_{CEO}.

Last, but not least, a major evolution appeared in the backend of line (BEOL), i.e., the metallization stack, with the introduction of thick copper module(s). It became standard starting from the B9MW technology [14], but an ultra-thick Cu option has been developed in B7RF [12]. The combination of ultra-thick via(s) and line(s) allows decreasing metal resistances and move away RF passives from the substrate, reducing substrate losses and so improving RF passive performances. The use of a high resistivity (HR) substrate does not always solve the limitation of standard resistivity (SR) substrate, as parasitic conduction layer can appear during the process. However, the HR substrate is appealing to integrate high-performance switches [19]. Benefits of RF optimized BEOL are extensively discussed in Section III.B

III. BICMOS055X: A DISRUPTIVE TECHNOLOGY IN ST ROADMAP

As shown in Table 1, the B55X technology gathers several innovations compared to the previous nodes. In addition, the technology has been defined to provide a versatile offer to serve multiple applications.

A. *An innovative SiGe HBT architecture*

Although the double-polysilicon self-aligned (DPSA) architecture using

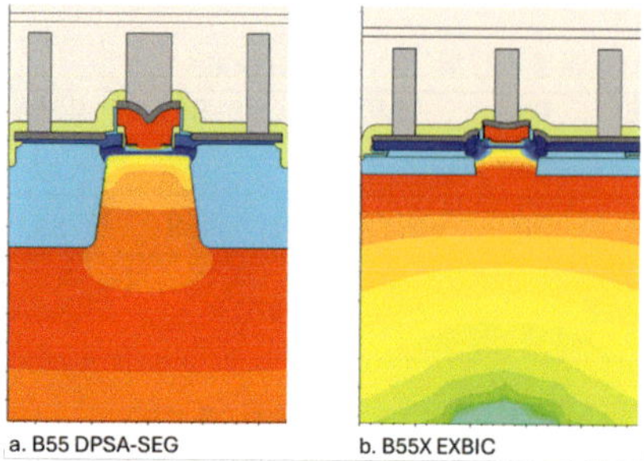

Fig. 2. TCAD cross-sections of the DPSA-SEG (a) and EXBIC (b) architectures used in STMicroelectronics B55 [15] and B55X [16][18] technologies, respectively.

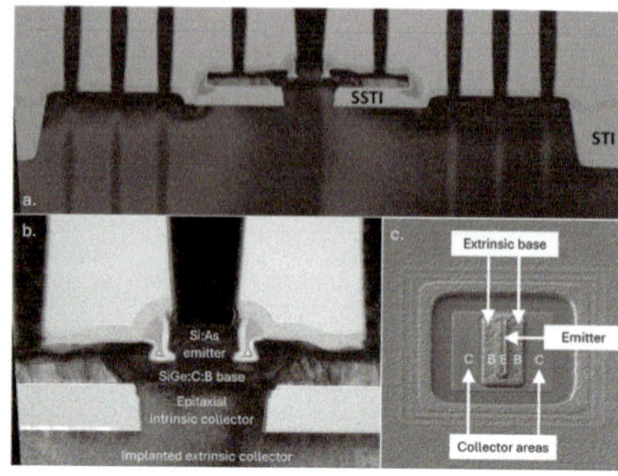

Fig. 3. B55X SiGe HBT (EXBIC architecture: TEM cross-sections at the end of fabrication of the whole transistor (a) and zoom-in of the emitter–base area (b) [18] and SEM top view after extrinsic base patterning (c).

an SEG of the base brought a significant improvement of R_B (thanks to the emitter–base self-alignment), further reducing this resistance collides with the way the link between the intrinsic base and the extrinsic base is made. Indeed, this link is done during the SEG of the base, leaving little room to optimize independently the intrinsic and extrinsic parts of R_B [20]. This issue is addressed with the last generations of architecture featuring a dedicated epitaxial growth of the extrinsic base [3]. Such a step is implemented in B55X [18] and combined with an innovative collector module featuring super shallow trench isolation (SSTI) and a shallow extrinsic collector layer, visible in Figures 2 and 3. It allows getting rid of DTI that are no longer required, neither to increase device compactness nor to reduce the collector–substrate (C_{CS}) capacitance [16]. An important point compared to similar collector architectures [3] is the use of standard transistor layout for which the emitter length of several micrometers can be drawn and the use of unit cells is not mandatory to increase the device area. This is illustrated by the CBEBC layout (i.e. 2 collector contacts "C" and 2 base contacts "B" on each side of the emitter contact "E") shown in Figure 3c. In addition, the height of the EXBIC architecture has been reduced compared to the DPSA-SEG architecture (cf. Figure 2), which is favorable for the emitter resistance R_E and opens its integration in a more advanced CMOS node [21]. f_T and f_{MAX} peak frequencies reported in Figure 1 are important KPIs, providing information on the maximum operation frequency and the gain that can be achieved at a given frequency. Another important KPI for low noise amplifier is the minimum noise figure (NF_{MIN}) that is known to correlate with f_{MAX} since it depends partly on same

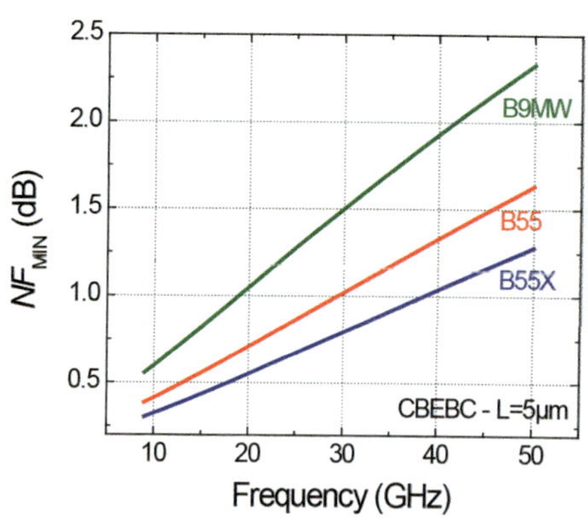

Fig. 4. Comparison of the evolution of SiGe HBT noise figure with the frequency (CBEBC transistor, V_{BE} = 0.82 V) between B9MW, B55, and B55X (data are from the models).

TABLE 2
B55X CMOS DEVICES OFFER

Gate oxide	Devices	Minimum L_G (nm)	nMOS	pMOS	RF model
5 nm	GO2 CMOS	280	SVT	SVT	Yes
1.8 nm	LP CMOS	60	LVT	LVT	Yes
			SVT	SVT	Yes
			HVT	HVT	No
	HPA CMOS	140	HPA	HPA	Yes
	LP SRAM	70	SVT	SVT	NA
			HBT	HBT	NA
1.3 nm	GP CMOS	45	LVT	LVT	Yes
			SVT	SVT	Yes
			HVT	HVT	No

parameters, a key one being R_B. Figure 4 compares the evolution of NF_{MIN} with the frequency between B55X and the two previous generations (B55 and B9MW). B55X SiGe HBT exhibits a strong reduction of the noise figure compared to previous generation with NF_{MIN} values below 0.5 and 1.0 dB in Ku and Ka bands, respectively. These values represent the current state-of-the-art [18].

B. Versatile CMOS and BEOL offers

Innovation in B55X does not only lie in SiGe HBT architecture but also in the definition of the technology content and its versatility to meet the PPACS (cf. Section I) of each application. The main changes compared to the previous generation are the ability to select a reduced list of CMOS devices and several BEOL stacks.

Table 2 presents the list of metal–oxide–semiconductor (MOS) devices available both in B55 and B55X. While the three families of MOS transistors (corresponding to the three-gate oxide thicknesses) could not be separated in B55 (only some device V_T flavors are optional), they can be selected separately in B55X. Four-gate oxide options are possible in B55X:

- 5-nm CMOS only;
- 1.8-nm CMOS only;
- 5-nm CMOS + 1.8-nm CMOS;
- 5-nm CMOS + 1.8-nm CMOS + 1.3-nm CMOS.

Process flow, and more especially the gate oxidations scheme, has been defined to insure a full compatibility between the options. RF model is available for most of the CMOS transistors.

Figure 5 presents the different metallization stacks developed for the B9MW, B55, and B55X technologies. They exhibit one or even two ultra-thick Cu layers of 2.3 or 3 μm combined with ultra-thick vias of 1.5 or 2.7 μm and feature different numbers of thin and thick metal layers. It allows addressing different trade-offs between the RF performance of inductors and transmission lines (related to the thick metal layers), the digital performance (related to the thin metal layers) and the process cost, i.e., overall the technology competitiveness. The Q-factors of single and differential ended inductors of 300 and 800 pH at 10, 20, and 40 GHz are reported in Figure 6. As expected, the thicker is the stack, the better is the quality factor and selecting the right stack is a

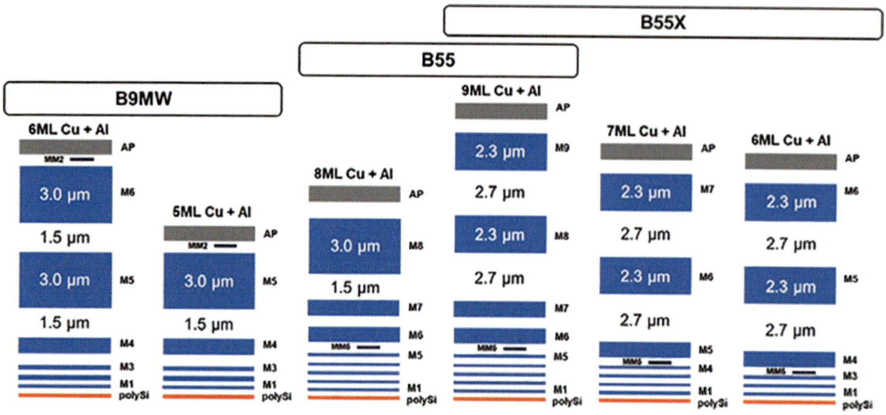

Fig. 5. Metallization stacks developed in the B9MW, B55, and B55X technologies.

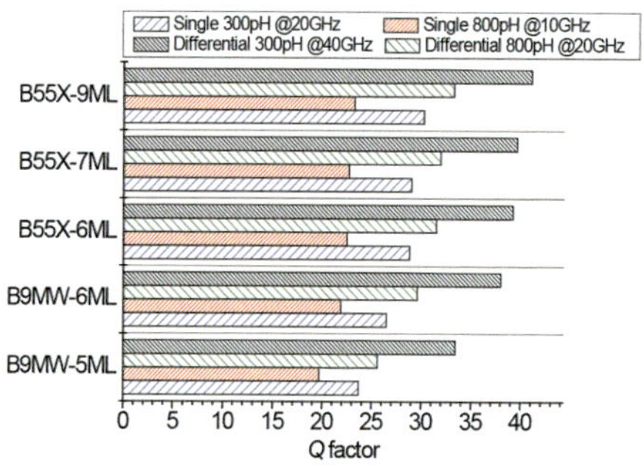

Fig. 6. Comparison of performance (Q-factor) of single ended and differential inductors of the BEOL stacks of Figure 5.

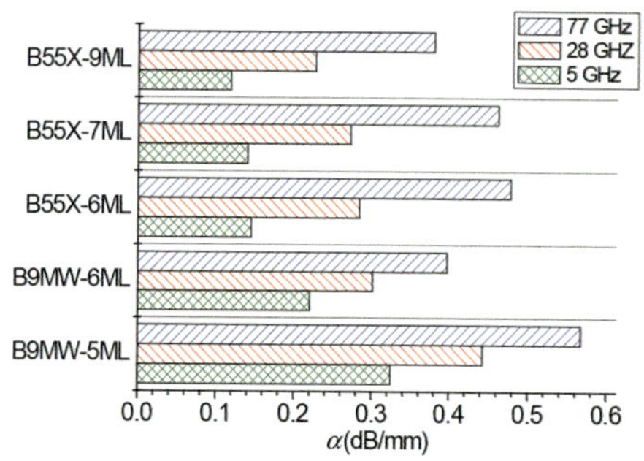

Fig. 7. Comparison of performance (attenuation constant α) of transmission lines of the BEOL stacks of Figure 5 (Z_c = 50W).

trade-off between wafer cost and circuit performance. However, it is interesting to note that the penalty of using a 5 Cu ML instead of 6 in B9MW is low at 10 GHz, while this difference is larger at 40 GHz. Also, performance is always better with B55X dual thick Cu stack, with a very low impact of the number of thin metal layers. Figure 7, showing the attenuation constant (α) of transmission lines at 5, 28, and 77 GHz for the same BEOL stacks, exhibits the same conclusions with respect to the importance of dual thick Cu module at higher frequencies and the low weight of the thin metal layers. Finally, although thin metal layers are thinner in B55X than in B9MW (in line with the digital density difference of the CMOS node), performances of inductors and transmission lines are always better (whatever the frequency) in B55X 6ML than in B9MW 6 ML. This is thanks to the thickening of the ultra-thick via + metal module in B55X (5 µm vs. 4.5 µm in B9MW), which represents a major breakthrough in the via thickness compared to previous generations.

III. BICMOS055X VERSUS THE COMPETITION

It is clear from previous sections that B55X features the best performance of the BiCMOS technologies developed at

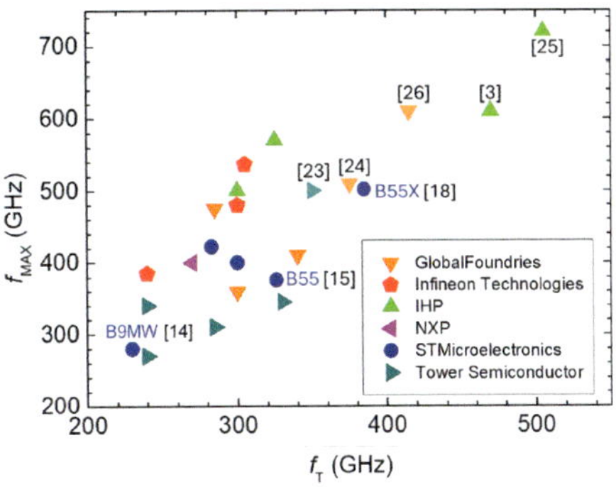

Fig. 8. Peak f_T and f_{MAX} values of high-speed SiGe HBT technologies (references are provided for f_T > 350 GHz and f_{MAX} > 400 GHz and ST technologies).

ST while taking care to adapt the technology offer to the needs by offering many options. However, its competitiveness can only be evaluated according to the competition. While there are only a few data published on metal stacks [22] and relative passive devices performance, SiGe HBT results are commonly reported. Figure 8 compares SiGe HBT f_T and f_{MAX} from different players with top performances (f_T > 350 GHz and f_{MAX} > 450 GHz) highlighted with a gray background. State-of-the-art technologies in production for about 10 years exhibit f_T and f_{MAX} of ~300 and ~400 GHz, respectively (± 50 GHz). B55X belongs to a new generation of technologies using an epitaxial base link aiming at dramatically reducing R_B, leading to f_{MAX} > 450 GHz that could not be achieved in previous generations. Looking at the area highlighted in gray, B55X [18] appears in the low f_T/f_{MAX} corner, very close to [23] and [24]. Best performance [25] has been demonstrated in a bipolar-only process, i.e., not compatible with CMOS, the CMOS compatibility being recovered in [3] for a 130-nm node (with an aluminum BEOL). CMOS node (and related substrate) is indeed an important difference between these "top" technologies. While [23] is based on 180-nm CMOS, [24] and [26] results are obtained on a 45-nm CMOS, built on partially depleted silicon-on-insulator substrate, which represents the current CMOS state-of-the-art in BiCMOS. Of course, it comes with a higher complexity/cost compared to 55-nm bulk CMOS.

IV. CONCLUSION

In the recent past years, SiGe BiCMOS development has been driven by the increase of the SiGe HBT RF performance on the one side and the move to more advanced CMOS nodes on the other side. The CMOS roadmap of these technologies was driven by some demanding applications, optical communications being the main one. While nanoscale nodes exhibit clear advantages [21], there is no high-volume application today calling for the integration of SiGe HBT with very advanced CMOS. On the contrary, the LEO Satcom application that drives the production volumes today requires a moderate CMOS density. In this context, a 55-nm node appears as a sweet spot regarding the CMOS performance and complexity. The combination of this CMOS node (with a versatile device offer), with state-of-the-art SiGe HBT and metallization stacks position B55X as a unique technology in the current BiCMOS landscape. This technology marks a shift in ST roadmap concerning the SiGe HBT architecture and device offer versatility [16]. Next generations will probably follow this trend, keeping an important focus on the balance between performance and complexity. Heterogeneous integration will probably provide interesting perspectives in this respect.

ACKNOWLEDGMENT

The authors thank all the Process Development, Design and Technology and Development Platform teams from STMicroelectronics for their contributions to B55X. They also thank N. Gambetta for the complementary information on 0.25-µm BiCMOS technologies.

References

[1] E. Preisler, "A Commercial Foundry Perspective of SiGe BiCMOS Process Technologies," in *Proc. IEEE BCICTS*, Virtual Event, 2020.

[2] P. Chevalier, W. Liebl, H. Rücker, A. Gauthier, D. Manger, B. Heinemann, G. Avenier, J. Böck, "SiGe BiCMOS Current Status and Future Trends in Europe," in *Proc. IEEE BCICTS*, San Diego, CA, USA, October, 2018, pp. 64–71.

[3] H. Rücker and B. Heinemann, "Device Architectures for High-Speed SiGe HBTs," in *Proc. IEEE BCICTS*, Nashville, TN, USA, November, 2019.

[4] M. Schröter, T. Rosenbaum, P. Chevalier, B. Heinemann, S.P. Voinigescu, Member IEEE, E. Preisler, J. Böck, A. Mukherjee, "SiGe HBT Technology: Future Trends and TCAD-Based Roadmap," in *Proceedings of the IEEE*, Vol. 105, No. 6, pp. 1068–1086, June 2017.

[5] P. Chevalier, "SiGe Speaks to the Sky," *International Microwave Symposium*, Connected Future Summit (CFS-8), San Diego, June 2023.

[6] "The Evolving Role of Optics in AI Clusters," Lightcounting, January 2024, [Online]. Available: https://www.lightcounting.com/newsletter/en/january-2024-optics-for-ai-clusters-308

[7] S. Nicoleau, J.-L. Champseix, D. Tagarian, F. Boeuf and P. Quinio, "Developing Sustainable Technologies for a more Sustainable Future," in *Proc. IEDM*, San Francisco, CA, USA, 2023.

[8] Sustainable Technologies Enabling Future Telecom Applications, [Online]. Available: https://www.shiftkdt.eu/

[9] Move2THz, Sustainable Indium Phosphide (InP) platform and ecosystem upscaling, enabling future mass market (sub)-THz applications, [Online]. Available: https://www.move2thz.eu/

[10] A. Monroy, M. Laurens, M. Marty, D. Dutartre, D. Gloria, J.L. Carbonero, A. Perrotin, M. Roche, A. Chantre, "BiCMOS6G: A high performance 0.35 mm SiGe BiCMOS technology for wireless applications," in *Proc. IEEE BCTM*, Minneapolis, MN, USA, October 1999, pp. 121–124.

[11] H. Baudry, B. Martinet, C. Fellous, O. Kermarrec, Y. Campidelli, M. Laurens, M. Marty, J. Mourier, G. Troillard, A. Monroy, D. Dutartre, D. Bensahel, G. Vincent, A. Chantre, "High performance 0.25 µm SiGe and SiGe:C HBTs using non selective epitaxy," in *Proc. IEEE BCTM*, Minneapolis, MN, USA, October 2001, pp. 52–55.

[12] H. Baudry, B. Szelag, F. Deleglise, M. Laurens, J. Mourier, F. Saguin, G. Troillard, A. Chantre, A. Monroy, "BiCMOS7RF: a highly manufacturable 0.25 µm BiCMOS RF-applications-dedicated technology using nonselective SiGe:C epitaxy," in *Proc. IEEE BCTM*, Toulouse, France, September 2003, pp. 207–210.

[13] M. Laurens, B. Martinet, O. Kermarrec, Y. Campidelli, F. Deleglise, D. Dutarte, G. Troillard, D. Gloria, J. Bonnouvrier, R. Beerkens, V. Rousset, F. Leverd, A. Chantre, A. Monroy, "A 150 GHz f_T / f_{MAX} 0.13 µm SiGe:C Technology," in *Proc. IEEE BCTM*, Toulouse, France, September 2003, pp. 199–202.

[14] G. Avenier, M. Diop, P. Chevalier, G. Troillard, N. Loubet, J. Bouvier, L. Depoyan, N. Derrier, M. Buczko, C. Leyris, S. Boret, S. Montusclat, A. Margain, S. Pruvost, S. T. Nicolson, K. H. K. Yau, N. Revil, D. Gloria, D. Dutartre, S. P. Voinigescu, A. Chantre, "0.13 µm SiGe BiCMOS Technology Fully Dedicated to mm-Wave Applications," *IEEE Journal of Solid-State Circuits*, Vol. 44, No. 9, September 2009, pp. 2312–2321.

[15] P. Chevalier P. Chevalier, G. Avenier, G. Ribes, A. Montagné, E. Canderle,

D. Céli, N. Derrier, C. Deglise, C. Durand, T. Quémerais, M. Buczko, D. Gloria, O. Robin, S. Petitdidier, Y. Campidelli, F. Abbate, M. Gros-Jean, L. Berthier, J.D. Chapon, F. Leverd, C. Jenny, C. Richard, O. Gourhant, C. De-Buttet, R. Beneyton, P. Maury, S. Joblot, L. Favennec, M. Guillermet, P. Brun, K. Courouble, K. Haxaire, G. Imbert, E. Gourvest, J. Cossalter, O. Saxod, C. Tavernier, F. Foussadier, B. Ramadout, R. Bianchini, C. Julien, D. Ney, J. Rosa, S. Haendler, Y. Carminati, B. Borot, "A 55 nm Triple Gate Oxide 9 Metal Layers SiGe BiCMOS Technology Featuring 320 GHz f_T / 370 GHz f_{MAX} HBT and High-Q Millimeter-Wave Passives", in *Proc. IEEE IEDM*, San Francisco, CA, USA, 2014, pp. 77–79.

[16] P. Chevalier, F. Cacho, C. Durand, N. Derrier, V. Milon, F. Monsieur, A. Gauthier, H. Audouin, M. Buczko, D. Céli, C. Deglise-Favre, C. Diouf, O. Foissey, M. Hello, N. Guitard, S. Madassamy, C. Renard, S. Ramirez-Ruiz, F. Sonnerat, V. Yon, D. Gloria, G. Waltisperger, "A Versatile 55-nm SiGe BiCMOS Technology for Wired, Wireless, and Satcom Applications," Accepted to *IEEE BCICTS*, Fort Lauderdale, FL, USA, 2024.

[17] P. Chevalier, T.F. Meister, B. Heinemann, S. Van Huylenbroeck, W. Liebl, A. Fox, A. Sibaja-Hernandez, A. Chantre, "Towards THz SiGe HBTs," in *Proc. IEEE BCTM*, Atlanta (GA), USA, October 2011, pp. 57–65.

[18] A. Gauthier, E. Brezza, A. Montagné, N. Guitard, J. Azevedo Goncalves, M. Buczko, S. Jan, N. Derrier, D. Céli, C. Deglise-Favre, J. Ma, H. Audouin, F. Deprat, D. Ristoiu, L. Berthier, L. Clément, B. Grelaud, C. Rougier, P. Chevalier, "Low-noise Si/SiGe HBT for LEO satellite user terminals in Ku-Ka bands," in *Proc. IEEE BCICTS*, Monterey, CA, USA, October 2023.

[19] F. Gianesello, C. Charbuillet, N. Derrier, D. Muller, C. Diouf, D. Ney, C. Deglise-Favre, I. Sicard, M. Ali Nsibi, R. Debroucke, M. Buczko, C.A. Legrand, P. Cathelin, F. Paillardet, J.C. Mas, P. Chevalier, D. Gloria, "0.13 µm HR SiGe BiCMOS Technology exhibiting 169 fs Ron x Coff Switch Performance targeting WiFi 6E Fully-Integrated RF Front-End-IC Solutions," in *Proc. IEEE IEDM*, San Francisco, CA, USA, December 2022, pp. 11.7.1–11.7.4.

[20] E. Canderle, P. Chevalier, A. Montagné, L. Moynet, G. Avenier, P. Boulenc, M. Buczko, Y. Carminati, J. Rosa, C. Gaquière, A. Chantre, "Extrinsic Base Resistance Optimization in DPSA-SEG SiGe:C HBTs," in *Proc. IEEE BCTM*, Portland, OR, USA, November 2012, pp. 149–152.

[21] P. Chevalier, G. Avenier, E. Canderle, A. Montagné, G. Ribes, V.T. Vu, "Nanoscale SiGe BiCMOS technologies: From 55 nm reality to 14 nm opportunities and challenges," in *Proc. IEEE BCTM*, Boston, MA, USA, October 2015, p 80–87.

[22] P. Chevalier and A. Pallotta, "Silicon Germanium spins cellular and space communication networks," Accepted for publication in *IEEE Microwave Magazine*.

[23] S. Phillips, E. Preisler, J. Zheng, S. Chaudhry, M. Racanelli, M. Müller, M. Schröter, W. McArthur, D. Howard, "Advances in foundry SiGe HBT BiCMOS processes through modeling and device scaling for ultra-high speed applications," in *Proc. IEEE BCICTS*, Virtual Event.

[24] J.J Pekarik, V. Jain, C. Kenney, J. Holt, S. Khokale, S. Saroop, J.B. Johnson, K. Stein, V. Ontalus, C. Durcan, M. Nafari, T. Nesheiwat, S. Saudari, E. Yarmoghaddam, S. Chaurasia, A. Joseph, "SiGe HBTs with f_T/f_{MAX} ~ 375/510GHz Integrated in 45nm PDSOI CMOS," in *Proc. IEEE BCICTS*, Virtual Event.

[25] B. Heinemann, H. Rücker, R. Barth, F. Bärwolf, J. Drews, G. G. Fischer, A. Fox, O. Fursenko, T. Grabolla, F. Herzel, J. Katzer, J. Korn, A. Krüger, P. Kulse, T. Lenke, M. Lisker, S. Marschmeyer, A. Scheit, D. Schmidt, J. Schmidt, M. A. Schubert, A. Trusch, C. Wipf,

D. Wolansky, "SiGe HBT with f_t/f_{max} of 505 GHz/720 GHz," in *Proc IEEE IEDM*, San Francisco, CA, USA, December 2016, pp. 51–54.

[26] V. Jain, J.J. Pekarik, C. Kenney, J. Holt, C. Durcan, J.B. Johnson, S. Saroop, M. Nafari, V. Ruparelia, S.K. Gedela, P.K. Sharma, V. Ontalus, S. Khokale, S. Chaurasia, V. Vanukuru, A. Joseph "415/610GHz f_T/f_{MAX} SiGe HBTs Integrated in a 45nm PDSOI BiCMOS process," in *Proc IEEE IEDM*, San Francisco, CA, USA, December 2022, pp. 266–269.

Pascal Chevalier (Member, IEEE) received the Ph.D. degree in electronics from the University of Lille, France, in 1998 for his work on InP-based HEMT. He joined Alcatel Microelectronics, Belgium, in 1999, where he contributed to the start of RF BiCMOS. Since joining STMicroelectronics, Crolles, France, in 2002, he has been working on the development of SiGe BiCMOS and RF-SOI CMOS technologies and related devices, with a long-lasting research interest in SiGe HBT. He led the RF-SOI CMOS, BiCMOS & Photonics Technologies R&D team. He is currently Analog/RF Technology Architect and is a Fellow of Technical Staff. Dr. Chevalier has authored or co-authored over 200 technical journal papers and conference publications. He has served on the Technical Program Committees of the IEEE BCTM, the ECS SiGe Symposium, the IEEE BCICTS, and the IEEE IEDM conferences. He has been a member of the RF & AMS Technologies section of the ITRS, of which he led the Silicon Bipolar & BiCMOS subgroup.

Alexis Gauthier received the Ph.D. degree in electronics from the University of Lille, France, in 2019 for his work on new Si/SiGe heterojunction bipolar transistor architectures. He joined STMicroelectronics, Crolles, France, in 2016, where he is contributing to the development of the new BiCMOS technologies with a dedicated focus on Si/SiGe HBT. He is currently leading the development of 300-mm wafers RF BiCMOS technologies. Dr. Gauthier has authored or co-authored 10 technical journal papers or conference publications and is inventor or co-inventor of 18 granted patents.

Nicolas Guitard received the Ph.D. degree in microelectronic from the University of Toulouse, France, at LAAS/CNRS, in 2006. He studied the capability of low noise characterization to detect latent defects in ICs due to ESD stresses. From 2006 to 2009, he worked at NXP semiconductors (in the Netherlands) as an ESD/LU Engineer. In 2009, he joined STMicroelectronics (ST) in Crolles (France), where he developed ESD protection IO libraries in most advanced CMOS technologies and drives qualification activities for both ESD and latch-up. He has also driven several R&D activities in the field of EOS and advanced ESD concepts, leading to several international publications. Since 2016, Nicolas has joined TCAD team at ST Crolles. He is now involved in device and process simulations for CMOS & BiCMOS technologies to support ST technologies R&D.

Victor Milon received the Master's and the Engineering degrees in microelectronics with a specialization in analog and digital design at Grenoble INP PHELMA, UGA France, in 2017. After a first experience in defining design rules based on process variability, he is now a senior engineer responsible of CMOS and BEOL devices in 300-mm BiCMOS R&D Technology Platform Development team.

Frédéric Monsieur, photograph and biography not available at the time of publication.

Nicolas Derrier received the Engineering degree in computer sciences and scientific computation from Polytech school in Sophia-Antipolis, France, in 1999. He then received the Engineering Master degree in numerical and solid state physics at the school, Mines ParisTech, France, in 2000. He joined the group STMicroelectronics in 2000 as a software developer for Digital Place & Route Design flow. He then moved to Metallization Backend of Lines modeling activity. In 2008, he joined the SPICE Modeling group. Since 2008, he has been a modeling engineer for different compact models and different families of devices or technologies, with a strong specialty for BJT and HBT bipolars. He has been involved in different R&D BiCMOS technology developments and is a trainer for BiCMOS devices and models. Since 2022, he has also been a Senior Member of the Technical Staff of STMicroelectronics, for RF bipolar device characterization and HICUM compact model.

Claire Deglise-Favre, photograph and biography not available at the time of publication.

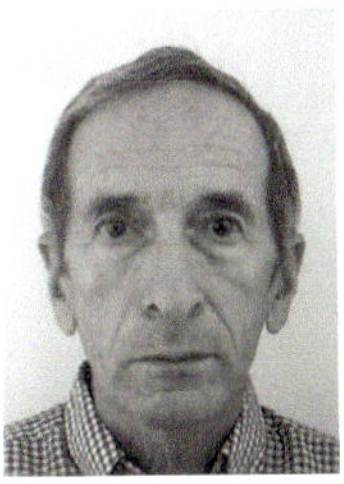

Didier Céli received the M.Sc. degree in electrical engineering from the Ecole Supérieure d'Electricité, Gif-sur-Yvette, France, in 1981. In 1982, he was with the Semiconductor Division, Thomson-CSF. He is currently with the Central R&D, STMicroelectronics, Crolles France, where he has worked in the field of modeling for advanced bipolar and BiCMOS technologies, including model development, parameter measurement, and extraction tools. He is now part of the Model Sustaining Group of STMicroelectronics as a technical expert. He also gives lectures and training courses on compact device modeling in engineering schools. He has authored or co-authored several technical papers related to his research. Didier Céli was a member of the CAD/Modeling Subcommittee of IEEE Bipolar/BiCMOS and Technology Meeting from 1998 to 2004. Since 2012, he has been in charge of the HICUM subcommittee of the CMC (compact model coalition).

Cédric Durand received the Master and the Engineering degrees in material sciences with a specialization in microelectronics at the University Joseph Fourier, Grenoble, France, in 2005. He then received the Ph.D. degree in 2009 from the University of Science and Technologies, Lille, France, made in collaboration with IEMN Lab at Villeneuve d'Ascq, France and CEA-LETI at Grenoble, France. The Ph.D. was dealing with RF MEMS and NEMS, particularly concerning the integration of electromechanical resonators. From 2009 to 2017, he has been involved on an R&D team concerning the development of integrated RF and photonics passive devices within STMicroelectronics. Since 2018, he has been acting as a technical leader R&D and technical expert within STMicroelectronics on the domain of RF and photonic components and associated packages. He is also a Senior Member of the Technical Staff of STMicroelectronics.

Ophélie Foissey, MScs, graduated in microelectronics and telecom in 2019 at PHELMA, Physics Electronics and Materials Engineering School. She firstly joined STMicroelectronics in 2016 to work on the development of passives and photonics components. She also worked on the development of PIN diodes integrated on silicon technology for advanced RF functions (switch, attenuator). Largely involved in the development of electromagnetic passive devices such as inductors, transformers, and couplers, she is also in charge of their modeling. In the same way, she contributes to RF and photonics circuits design, on top of customer support.

Florence Sonnerat received the Ph.D. degree in electrical engineering from Telecom Bretagne, Brest, France, in 2013. Until mid-2015, she worked in CEA-LETI, France, where she was in charge of RF passive component characterization. She then worked for Radiall, Voreppe, France for one year and designed RF connectors for high data rates applications. Since 2016, she works for STMicroelectronics, Crolles, France, on RF passive components and electromagnetic devices modeling. She authored and co-authored about 15 refereed journal and conference technical articles.

Fred Gianesello received the B.S. and M.S. degrees in electronics engineering from Institut National Polytechnique de Grenoble, Grenoble, France, in 2003 and the Ph.D. degree in electrical engineering from the Joseph Fourier University, Grenoble, France, in 2006. Dr. Gianesello has authored and co-authored more than 200+ refereed journals and conference technical articles. He has served the TPC of IEEE International SOI Conference, IEEE International Solid-State Circuits Conference, IEEE International Electron Device Meeting Conference, and IEEE EuMIC Conference and is currently serving the TPC of IEEE RFIC. Dr. Gianesello has been working for ST on the development of silicon-based RF and millimeter wave technologies during the past 20 years targeting RF and mmW transceivers and front-end module products. He is currently working for STMicroelectronics in Crolles, France, as technical expert providing support to customer and performing business prospection for RF, mmW, and optical products.

Daniel Gloria received in 1995 the engineering degree in electronics from the Ecole Nationale Supérieure d'Electronique et de Radioélectricité and the M.S.E.E. in optics, optoelectronics and microwaves design systems from the Institut National de Grenoble (INPG).

He spent two years, from 1995 to 1997, in ALCATEL Bell Network System Labs, in Charleroi, Belgium, as an RF designer engineer and was involved in the development of the cablephone RF front end and its integration in hybrid-fiber-coax telecommunication networks. Since 1997, he has been working within STMicroelectronics, in the Technology Design Platform Department, Crolles. His research interests are in the development and optimization of RF and photonics platforms for applications in BiCMOS and CMOS advanced technologies.

RF Technology Roadmap for 5G and 6G RF Front-end Systems

Yvan Morandini

Abstract—**The advent of 5G advanced and transition to 6G wireless systems are ushering in a new era of challenges for radio frequency (RF) front-end module design and integration. To seamlessly navigate this technological shift, which incorporates new sub-20-GHz spectrum (FR3) allocation and the expanded utilization of millimeter-wave frequencies (FR2), the foundational RF technology substrate must be engineered to meet the stringent performance demands of next-generation 5G advanced and 6G RF front-ends. This evolution requires cutting-edge semiconductor technologies to optimize signal integrity, power efficiency, and system integration. In our forthcoming discussion, we will present RF substrate technology roadmap, crucial for architecting the resilient, high-performance RF systems that will underpin the next-generation 5G and 6G wireless infrastructure.**

In this paper, we will provide a roadmap of RF technology including the engineered substrates enabling to solve design challenges of wireless communication systems.

Index Terms— **Engineered substrates, RF front-end, RF-SOI, FD-SOI, POI, GaN and InP**

I. INTRODUCTION

THIS paper describes recent developments of engineered substrate for next-generation cellular RF front-end (RFFE) design, addressing the evolving hardware requirements for future mobile devices. The first section discusses the evolution and challenges of 5G and upcoming 6G RFFEs. The following sections present key advancements in engineered substrates aimed at overcoming these challenges, including RF-SOI (Radio Frequency Silicon on Insulator), FD-SOI (Fully Depleted Silicon on Insulator), piezo on insulator (POI), GaN, and InP technologies.

II. 5G AND NEXT GENERATION 6G RFFE EVOLUTION AND CHALLENGES

The key challenges and evolution of 5G RFFE are comprehensively detailed in [1]. The frequency bands for 5G new radio (5G NR) now include Wi-Fi bands at 2.4, 5.8, and 6 GHz. To support high data rates, the RFFE architecture has become increasingly complex, accommodating the coexistence of Wi-Fi and 5G, new frequency bands, and dual connectivity. Within the FR1 spectrum, new sub-6GHz bands have been introduced to address the data limitations of 4G. These bands, in conjunction with reframed 4G bands, require additional receive and transmit modules. In the FR2 spectrum, 5G millimeter-waves (mmWave) at 28 and 39 GHz offer significantly higher bandwidths compared to 4G, with bands extending to 52 and 71 GHz. The high frequency losses associated with these high frequency bands make it essential to integrate the antenna and RFFE into a

This work is related to Move2THz, a project supported by the Chips Joint Undertaking (Chips JU) under grant agreement number 101139842. The Chips JU receives support from the European Union's Horizon Europe research and innovation program and the National Authorities.

Yvan Morandini (yvan.morandini@soitec.com) is with Soitec, France.

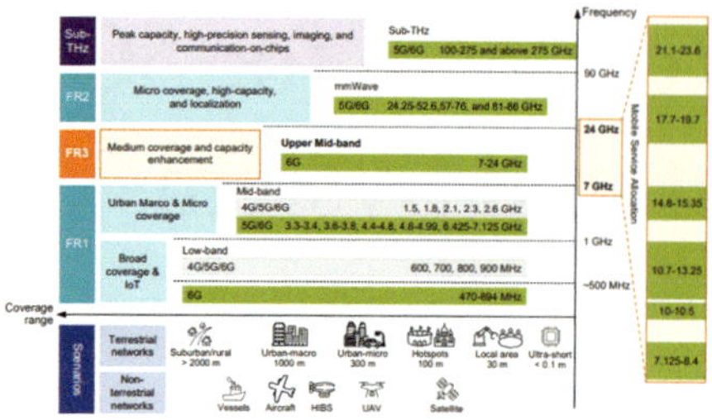

Fig. 1. Spectrum overview for mobile networks [2].

single module, referred to as antenna in package. Multiple antenna modules are required within devices to mitigate the effects of user hand interference. The 5G mobile architecture requires dual connectivity to utilize both 5G and LTE bands, as standardized in 3GPP Release 15, thereby improving coverage and data rates. To facilitate faster data transfer, 5G smartphones must support 4 × 4 MIMO for downlink and 2 × 2 MIMO for uplink. The increased complexity arising from coexistence of 4G and 5G, along with the introduction of new 5G bands, underscores the need for advanced RF front-end modules (FEMs).

Given the congestion in the sub-6G spectrum, the telecommunications industry is increasingly exploring higher frequencies to achieve greater bandwidth, which is essential for future 5G and 6G deployments. The newly allocated FR3 spectrum, ranging from 7 to 24 GHz (Figure 1), offers a promising solution by combining the broad area coverage characteristics of sub-6-GHz frequencies with the enhanced data rates and capacity typical to mmWave bands. FR3 provides distinct advantages over FR1 by offering wider channel bandwidths. Compared to FR2, FR3 experiences reduced signal attenuation and entails lower hardware implementation costs, making it a valuable option for next-generation networks.

To efficiently operate RFFEs within the 7–24 GHz range, novel transceivers and RFFEs must be designed. These components must incorporate up/down converters, ADCs (Analog to Digital Converter) with wider bandwidth considerations, and power amplifiers (PA) capable of linear output in FR3. Furthermore, existing transceiver architectures developed for FR1 and FR2 will need to be adopted for compatibility with FR3. FR3 represents an ideal band for hybrid beamforming architecture, balancing spatial multiplexing in FR1 while reducing beamformer complexity and power consumption in FR2. The central challenge lies in harmonizing optimization-based and learning-based designs for the hybrid transceiver architecture. As we advance this architecture, extending it toward RF-SOI technology will be essential for achieving optimal performance.

Utilizing the THz band presents significant challenges for transceiver hardware design [3]. Operating at such high frequencies imposes stringent requirements on semiconductor technology. Even with state-of-the-art technologies, the operating frequency may approach or even exceed the maximum frequency where the semiconductor can effectively provide power gain. This results in a severely degraded receiver noise figure and reduced transmitter efficiency compared to lower frequency operations. To maximize high-frequency gain, technology must employ scaled-down feature sizes, necessitating low supply voltages to maintain reliability. This reduction in supply voltage, however, diminishes the achievable transmitter output power. Additionally, the degraded receiver noise figure, reduced antenna aperture, and wide signal bandwidth at these high frequencies naturally leads to very short link distances.

The key challenges for this evolution are:

- Coexistence of 5G, 6G, and Wi-Fi bands: Increased interference risk due to overlapping frequency usage.

- New dedicated bands for 5G and 6G: More carrier aggregation will require additional filter paths, increasing the strain on the RF bill of materials footprint.
- Bandwidth requirements: Support for bandwidths up to 100 MHz.
- Higher power class devices: Devices requiring up to 26 dBm at the antenna, as defined by 3GPP. This new power class category establishes the maximum transmit power over the full 5G NR channel bandwidth, addressing the link budget limitation from user equipment to the base station. Higher transmit power from user equipment enables better cell coverage, necessitating more efficient high PAs.
- Improved PA efficiency: Specifically enhancing efficiency at higher frequencies.
- Technology scaling: Extend CMOS maximum frequency and III-V RF technology CMOS compatible.

III. NEXT-GENERATION RF-SOI TECHNOLOGIES

While RF CMOS SOI technologies are well-established for RF switches in FEMs, they have also been implemented for low-noise amplifiers (LNAs) by reducing gate length to improve gain and noise figures. The integration of more FEM functions into a single chip becomes increasingly crucial in the FR2 mmWave and FR3 ranges to minimize package-related losses. Addressing this integration challenge requires continuous improvement in figures of merit such as f_t (cutoff frequency) and f_{max} (maximum oscillation frequency).

RF-SOI technology, depending on the application and frequency bands, can be compatible with CMOS technology down to 40 nm, as illustrated in Figure 2. Next-generation RF-SOI technologies will

Application	Frequency band	Range (coverage)	RF-SOI Technology platform			
			180nm	130/90nm	65/55nm	45/40nm
Antenna tuners	sub-6GHz		✓	✓	✓	
	6 - 15GHz			✓	✓	
	mmWave					
Switch	sub-6GHz	short	✓	✓	✓	✓
		wide	✓	✓	✓	
	6 - 15GHz			✓	✓	✓
	mmWave				✓	✓
LNA	sub-6GHz	short	✓	✓	✓	✓
		wide			✓	✓
	6-15GHz			✓	✓	✓
	mmWave				✓	✓
mmW standalone* RFFE	mmWave				✓	✓
						✓

Fig. 2. Applicability of RF-SOI technology platforms.

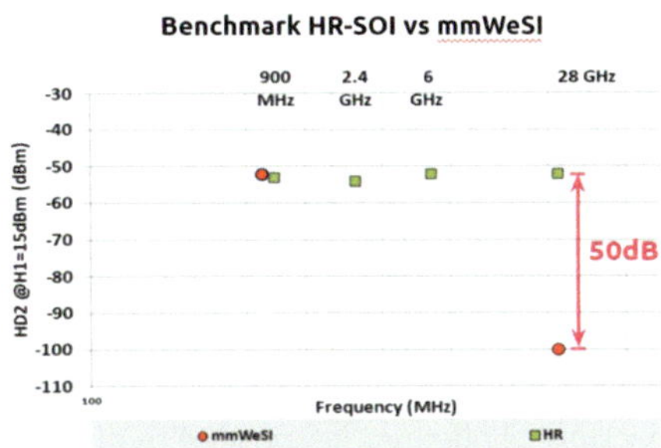

Fig. 3. Second harmonic as function of the frequency comparing HR-SOI with mmWave trap-rich RF-SOI. Input power: 15 dBm.

require to develop new RFeSI™ trap-rich technology, maintain good linearity performance, and be compatible with 40 nm CMOS technologies. Next-generation mmWave trap-rich substrate solutions will enable a lower second harmonic performance of 50 dBm while being compatible with stringent overlay performance requirements as depicted in Figure 3.

IV. NEXT-GENERATION FD-SOI TECHNOLOGIES

The use of a commercial low-leakage FD-SOI RF planar CMOS platform has successfully demonstrated the advantages of this mmWave RFFE architecture in the mobile market. Compared to CMOS bulk technology [4], it improves PA efficiency, enhances the noise figure of LNAs, and optimizes power consumption. The evolution of 5G advanced and 5G systems will require extending the maximum frequency of CMOS FD-SOI technology up to 500 GHz and looking into the integration of high resistivity option. Today, the 22 nm FD-SOI technology has demonstrated f_t/f_{max} of 370 GHz/410 GHz.

A promising solution to extend RF performance while enabling digital scaling is the use of strained layers. The Smartcut technique can be employed to transfer bi-axially strained silicon films, grown on fully relaxed SiGe buffer layers on Si bulk donor wafers, to create unique strained SOI. This approach leverages the carrier mobility enhancement offered by tensile-strained silicon [5,6].

The high resistivity substrate option is a significant enhancement to achieving ultimate mmWave performance. The engineering challenges associated with this new wafer generation are detailed in the study [7]. Among the various options for highly resistive silicon, low and high interstitial oxygen (Oi) materials are mostly used to achieve the desired resistivity performance on SOI handles. However, these materials are less compatible with FD-SOI technology, as they can be highly sensitive to slip-line issues or wafer deformation, it could lead to overlay errors. Additionally, high Oi substrates may encounter inspectability and co-integration issues due to the presence of crystal originated particles. Soitec has developed a specific process to tackle these challenges. They demonstrated how substrates were specifically engineered to achieve the appropriate resistivity targets around 1000 Ω.cm, ensuring stability at depth through various additional anneals and maintaining compatibility with customer processes. This process also ensures excellent mechanical performance, minimizing slip-line generation during fabrication processes, and avoiding overlay issues typically associated with slip-lines or excessive presence of bulk micro defects at depth, which can cause dislocations and plastic deformation. Furthermore, [8] describes that the combination of high resistivity substrate with PN interface passivation technique is particularly relevant for FD-SOI. By utilizing alternating regions of P- and N-type doping, the conductive interface is locally interrupted by induced depletion junctions. This chain-series combination significantly increases the overall substrate impedance (ρeff).

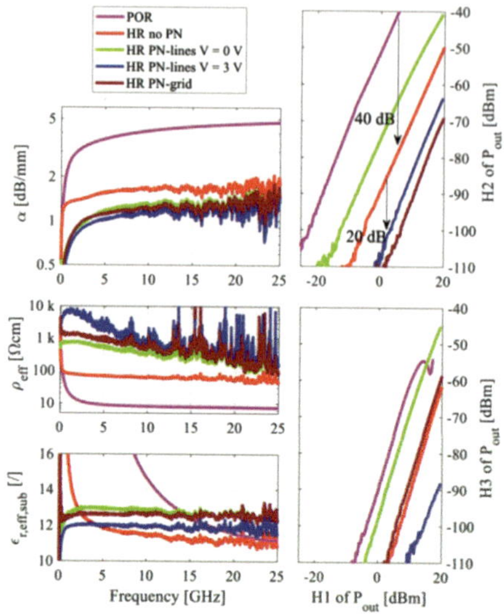

Fig. 4. RF substrate figures of merit were measured from M1M2 CPW lines. Small-signal data, including α (attenuation constant), ρeff (effective resistivity), and εr_{effsub} (effective relative permittivity of the substrate), were extracted using multiline thru-reflect-line calibration from multiple CPWs. Large-signal data, including harmonics H1, H2, and H3, were measured at a 900 MHz fundamental frequency on a 2-mm-long CPW [8].

Measurements of coplanar waveguides have demonstrated significant improvements in harmonic performance, as illustrated in Figure 4.

V. PIEZO MATERIAL FOR RF FILTER

The adoption of massive input massive output (MIMO) in receiver and

	SAW	TC-SAW	BAW	SAW on POI
Operating Frequency (GHz)	Low - High	Low – High	High Ultra High	Low - High Ultra High
Quality Factor (Bode Q) @2GHz	1500	<1500	4000 (AlN)	>4000
Coupling factor k^2	<7% ($LiTaO_3$)	<7%-$LiTaO_3$ <20% -$LiNbO_3$	<7% - AlN <12% - AlNSc	>8% - $LiTaO_3$ >20% -$LiNbO_3$
Temperature Compensation ($ppm.K^{-1}$)	>40 - $LiTaO_3$	<20 - $LiTaO_3$	>20	< 10 - $LiTaO_3$
Substrate complexity	Low	Low	Low	High
Process complexity	Low	Mid	High	Low
Integration	Low	Low	Low	High

Fig. 5. Piezo on insulator (POI) versus alternative RF filter technologies benchmarking.

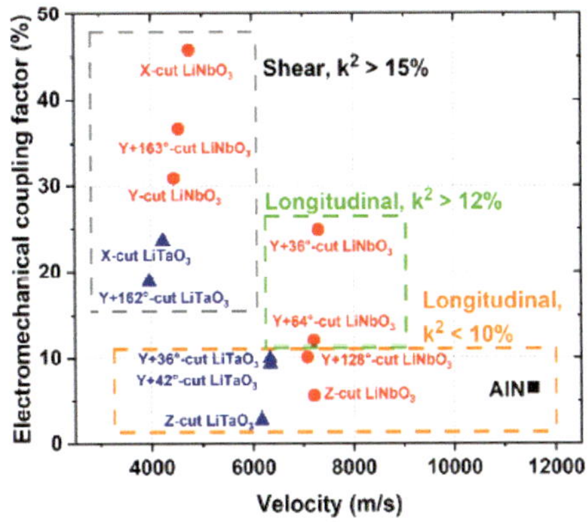

Fig. 6. Electromechanical coupling factors and bulk wave velocities for various $LiNbO_3$ and $LiTaO_3$ commercial cuts [9].

transmitter front-ends, combined with the increase in the number of bands, carrier combinations, and signal paths, significantly raises the demand for RF filtering elements in the RFFE. POI substrates present a promising solution to accommodate this growing demand within the same or even smaller footprint. These substrates provide stable high rejection and low loss even at high temperatures, and they simplify the manufacturing process.

Figure 5 illustrates the advantages of using POI substrates by comparing competitive technologies over different frequency bands.

In single crystal materials like $LiNbO_3$ and $LiTaO_3$, the polarization and velocity of acoustic waves can be adjusted by choosing the appropriate crystal orientation. These materials enable electromechanical coupling factors up to 45% and acoustic velocities between 3960 and 7320 m/s, which is depicted in Figure 6. This versatility makes them suitable for high-frequency applications.

The use of $LiTaO_3$ has been demonstrated for POI technology and preliminary studies are extending the use of $LiNbO_3$ beyond 3.5 GHz. Soitec and partners are investing $LiNbO_3$ as a good alternative to extend POI technology beyond 3.5 GHz and to extend the bandwidth of RF filter below 3.5 GHz.

VI. RF GAN TECHNOLOGY

RF GaN technology has shown significant advantages for RF PAs in 5G base stations, primarily utilizing GaN on SiC substrates. However, as massive MIMO infrastructure and beamforming transmission are deployed, the output power handled by individual PAs significantly decreases, despite a corresponding increase in the number of PA devices. Moreover, the adoption of the FR3 band for mobile user equipment could potentially disrupt the PA technology. Addressing the new device requirements will necessitate specific advancements in epitaxial GaN development [10] through:

- Reduction of conducting RF losses: Minimizing RF losses at the GaN/Si interface up to mmWave frequencies is crucial for optimal performance.
- Engineering low-trapping (Al,Ga)N buffer stacks: This involves incorporating carbon into resistive layers and back-barriers to enhance electron confinement and reduce trapping effects.

- Barrier design: Careful design of the barrier layer is necessary to ensure efficient electron mobility and overall device performance.
- Capping or passivation layer: Special attention is required for the capping or passivation layer, particularly when using MOCVD-grown SiN, to protect the device and maintain its electrical properties.

The study [11] describes the suitability capability of RF GaN for mobile user equipment design and highlighted advancements in epitaxial layer stack development. To meet voltage requirements, GaN device development incorporates a new epilayer featuring in undoped AlInN barrier layer, undoped GaN buffer layer, and an AlGaN back barrier on Si substrates. Devices using this structure exhibit exceptional performance across a voltage range from 1.5 to 12 V, achieving power density (Pd) values of 2.66 and 4.23 W/mm at Vds of 8 and 12 V, respectively, with power added efficiency (PAE) exceeding 60%.

Soitec [12] developed a pioneering concept with Smart Cut™ technology, introducing GaN-on-X substrates that overcome the physical limitations of heteroepitaxial growth. This method involves transferring a thin GaN film onto a handling substrate (such as Si, SOI, polySiC, etc.), tailored specifically for RF applications. This innovation enables the creation of devices with superior thermal resistance (Rth), higher power density, improved power efficiency, and reduced memory effects, enhancing linearity.

VII. INDIUM PHOSPHIDE (INP)

The sub-THz spectrum (FR4) including frequencies beyond 100 GHz, such as the D-band, gained significant interest for achieving the ultra-high data rate goals of next-generation 6G cellular networks. However, operating at carrier frequencies above 100 GHz presents substantial challenge, particularly in terms of power efficiency. To address these challenges, semiconductor and compound technologies with transistor f_{max} values exceeding 500 GHz may be necessary to enhance efficiency, gain, and noise performance. While SiGe technology holds the potential for improving silicon transistor performance, current advancements in InP technology offer the best front-end performance at sub-THz frequencies. This technology provides superior PA P_{out} and PAE as well as a low noise figure characteristics.

Despite its advantages, InP technology is limitedly used in markets with low volume production. That is mainly related to its manufacturing constraints, such as their small wafer diameters and high costs. To thrive in the consumer market, InP needs reliable and cost-effective manufacturing processes. An innovative method for producing large-diameter (up to 300 mm), cost-effective InP wafers is detailed in [10]. This method is based on wafer reconstruction and reclaiming. Building upon the foundational aspects of the SmartCut™ technology and the tiling approach, the process begins by selecting smaller diameter III-V wafers, such as those made from GaAs or InP. These are traditionally limited to 100 or 150 mm due to the constraints in bulk material production. These smaller wafers are then meticulously arranged and bonded onto a larger, 200 mm silicon wafer, creating a composite or pseudo-donor wafer. This approach not only circumvents the diameter limitations of traditional III-V substrates but also leverages the robustness and scalability of silicon infrastructure.

References

[1] Y. Morandini, "RF technology roadmap for 5G and 6G RF front end systems", ESSDERC 2023.

[2] Z. Cui,"6G Wireless Communications in 7-24 GHz Band: Opportunities, Techniques, and Challenges", IEEE System and control 2024

[3] H. Tataria "6G Wireless Systems: Vision, Requirements, Challenges, Insights, and Opportunities", Proceedings of the IEEE Volume: 109, Issue: 7, July 2021

[4] V. Bhagavatula et al., "A 5G FR2 Power-Amplifier With an Integrated Power-Detector for Closed-Loop EIRP Control," in IEEE Journal of Solid-State Circuits, vol. 57, no. 5, pp. 1257–1266, May 2022.

[5] C. Sun et al, "Enabling UTBB Strained SOI Platform for Co-integration of Logic and RF: Implant-Induced Strain Relaxation and Comb-like Device Architecture", VLSI 2020

[6] B. De Salvo et al, "A mobility enhancement strategy for sub 14nm power-efficient FDSOI technologies" IEDM 2014

[7] I. Bertrand, "Development of High Resistivity FD-SOI Substrates for mmWave Applications", ECS Transactions, Volume 108, Number 5, 2022

[8] M. Rack et al., "High-Resistivity Substrates with PN Interface Passivation in 22 nm FD-SOI", VLSI 2022

[9] M. Bousquet et al., "4.2 GHz LiNbO3 Film Bulk Acoustic Resonator,"

[10] 2021 IEEE International Ultrasonics Symposium

[11] M. Germain et al. "GaN epitaxial solutions for RF and Power devices"

[12] K. Takeuchi et al., "Low-Voltage Operation AlInN/GaN HEMTs on Si with High Output Power at sub-6 GHz", IMS June 2023

[13] G.Gaudin, ICNS 14, 14th International Conference on Nitride Semiconductors, 2023, Fukuoka, Japan, ThP-GR-LN4

[14] B. Ghyselen et al., "Large-Diameter III–V on Si Substrates by the Smart Cut Process: The 200 mm InP Film on Si Substrate Example," Phys. Status Solidi A, Vol. 219, Issue 4, Feb. 2022.

Yvan Morandini received the MSc degree in engineering from Grenoble Alpes University, France, in electronics and radiofrequencies in 2005 and the Ph.D. degree from Lille University in 2008. After three years in STMicroelectronics, he then moved to IBM, Dolphin Integration and then to SOITEC. He is currently Strategic Marketing Senior Manager at SOITEC. He has 20 years of experience in the semiconductor industry including characterization, modeling, and design. He has authored/co-authored over 15 papers in international, peer-reviewed journals and conferences, and is a Member of Institute of Electrical and Electronics Engineers (IEEE).

Advanced Substrate Technologies for Sub-THz Era

François Brunier

Abstract—**Electronic devices evolved significantly, driving the digital transformation toward a connected society. The increasing demand for performance, speed, and efficiency is pushing wireless applications to operate at sub-THz frequencies and beyond. Innovative substrates, such as fully depleted silicon on insulator (FD-SOI), indium phosphide on silicon (InPoSi), and SmartGaN, offer disruptive yet commercially viable solutions to efficiently utilize these frequencies. Achieving European leadership in these key markets requires establishing a complete and reliable supply chain from materials to applications. A so-called value chain model is applied in European KDT JU programs, which serves as a fast track to accelerate co-innovation and market adoption. This paper illustrates the innovation dynamics of the FD-SOI ecosystem in radio frequency applications up to 120 GHz, realized in the BEYOND5 project. It also explores technology disruption in the sub-THz frequency range by using InPoSi substrates in combination with the Smart Cut process. The Move2THz project aims to develop such technologies and demonstrate that it overcomes historical limitations of bulk InP. Finally, the paper demonstrates how the industry-proven Smart Cut technology can be adapted to other materials, such as gallium nitride, to complete the technological offering and meet market needs.**

Index Terms—**B5G, 6G, FD-SOI, mmWave, sub-THz, InP, InPoSi, Smart Cut, SmartGaN**

This work is supported by French funding under the Important Project of Common European Interest for Microelectronics and Communication Technologies (IPCEI ME/CT).

This work is also supported by the Chips Joint Undertaking (Chips JU) under grant agreement number 876124 (project BEYOND5) and under grant agreement number 10113988 (project Move2THz). The Chips JU receives support from the European Union's Horizon Europe research and innovation program and the National Authorities.

François Brunier (francois.brunier@soitec.com) is with Soitec, France.

I. INTRODUCTION ON ADVANCED SUBSTRATES

IN collaboration with the Leti (French Laboratoire d'électronique et des technologies de l'information) Soitec developed the patented Smart Cut process over 30 years ago. This process, illustrated in Figure 1, enables to manufacture advanced substrates for microelectronics by transferring a thin active layer of silicon or another semiconductor material from a donor substrate onto a second substrate that acts as a support. This Smart Cut process ensures high yield and surface quality for industrial manufacturability. It facilitates scaling up wafer size and volume to align with complementary metal–oxide–semiconductor (CMOS) manufacturing capacities, minimizes the use of rare resources, reduces the technology's ecological footprint and cost, and offers a wide range of possibilities for material integration, as illustrated in Figure 2. The combination of an adapted active layer on a functional support substrate creates a more resilient and less dependent value chain.

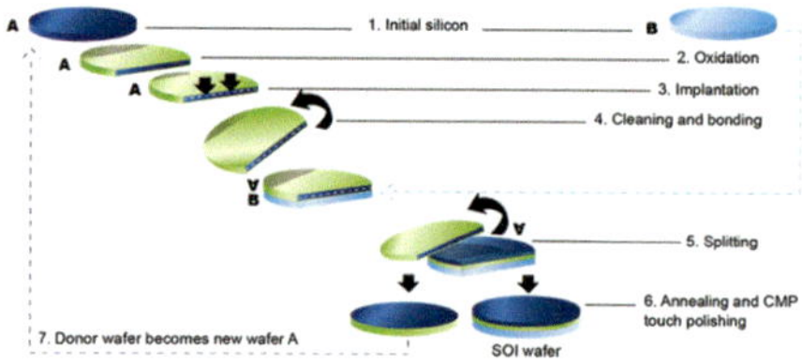

Fig. 1. Schematic illustration of the Smart Cut technology that manufactures SOI substrates.

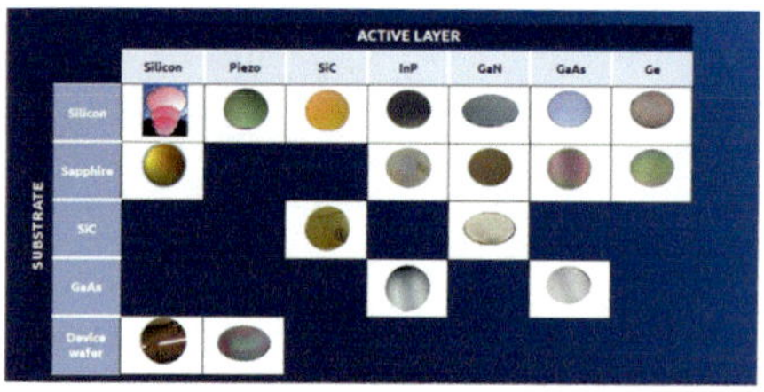

Fig. 2. Smart Cut technology combining an adapted active layer on a functional support substrate for a less dependent value chain.

II. RF TECHNOLOGY ON SOI – AN UNRIVALLED PPAC FOR 5G AND BEYOND

Radio frequency silicon on insulator (RF-SOI) and fully depleted silicon on insulator (FD-SOI) technologies based on SOI substrates meet the global 5G smartphone market requirements and offer new solutions for applications using the 5G frequency spectrum, ranging from low GHz bands to millimeter wave (mmWave) bands, including future prospects beyond 100 GHz. With its smaller device dimensions, FD-SOI enables very high cut-off frequencies, allowing circuit operations at very high frequencies, covering mmWave and beyond 100 GHz. The radio frequency front-end mmWave architecture, using a commercial low-leakage FD-SOI RF planar CMOS platform, has been successfully demonstrated in mobile market [1]. Additionally, FD-SOI allows co-integration of analog and digital capabilities into system-on-chip designs. Thanks to its back-biasing capability, FD-SOI also offers dynamic power/performance ratio tuning mechanisms. This makes FD-SOI also essential for automotive radar and Advanced Driver Assistance Systems (ADAS).

The KDT IA project BEYOND5, "Building the fully European supply chain on RF-SOI, enabling New RF Domains for sensing, Communication, 5G and beyond" leverages the RF-SOI and FD-SOI technologies to build a solid European ecosystem over different application domains. Started in June 2020, the project unites 39 European partners from 10 countries, covering the entire value chain with two SOI technologies (RF-SOI in STMicroelectronics, 22FDX in GlobalFoundries) and seven demonstrators. On the FD-SOI pilot line, GlobalFoundries develops and integrates new RF functions and reliability methodologies onto its 22FDX baseline. This includes the utilization of high resistivity base wafers to enhance the properties of 5G demonstrators and meet future mmWave system requirements.

This project's goal is to showcase these technology platforms at the system level. Figure 3 illustrates the consortium partners mapped onto the value chain, and Figure 4 depicts which technology platforms are incorporated in the different application demonstrators.

Each of the seven demonstrators selected the most competitive SOI technology based on its specific requirements,

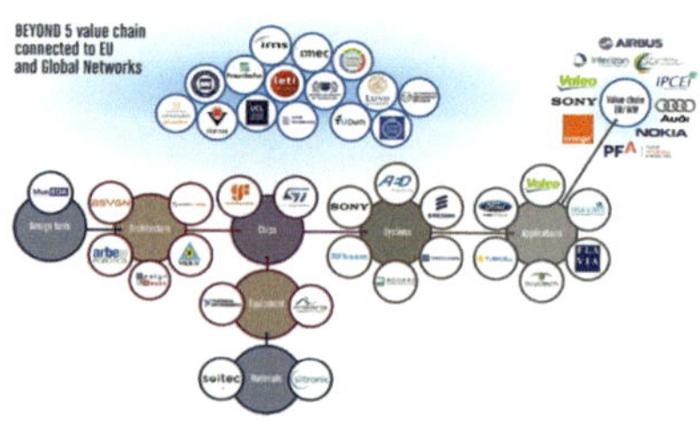

Fig. 3. BEYOND5 project consortium partners mapped onto the value chain.

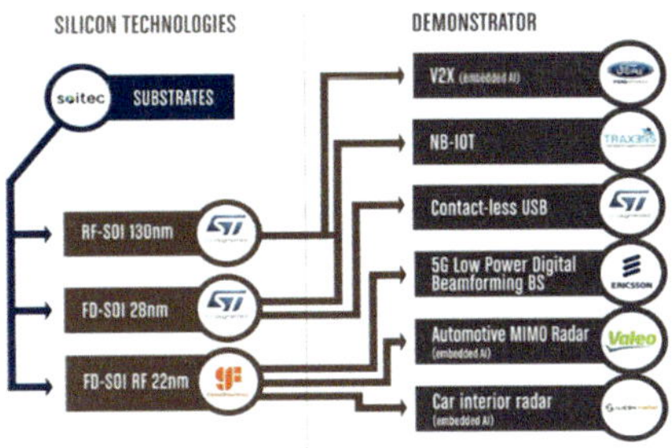

Fig. 4. BEYOND5 project demonstrators incorporate the different RF-SOI and FD-SOI technology flavor.

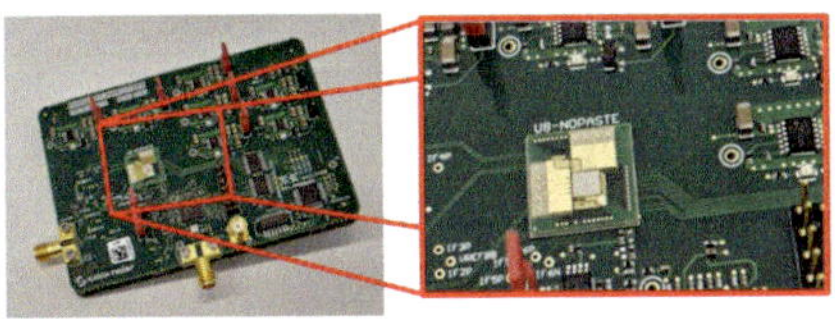

Fig. 5. 4 × 4 MIMO TRX 22FDX chip for the in-cabin radar demonstrator developed in the BEYOND5 project.

with key criteria including frequency range, performance, integration, power efficiency, and cost.

For the 122 GHz in-cabin radar and gesture recognition demonstrator, for example, the 22FDX was chosen for its RF performance, enabling low-power designs operating beyond 100 GHz. Its scaling effects facilitate true low-cost designs with minimal area consumption for processing units in system-in-package solutions. The in-cabin demonstrator, depicted in Figure 5, utilizes a 4 × 4 Multiple-Input Multiple-Output (MIMO) transceiver (TRX) chip with an in-package antenna operating in the 116–123 GHz band.

The block diagram of this millimeter-wave integrated circuit (MMIC) is illustrated in Figure 6. This TRX is developed using 22FDX technology from GlobalFoundries.

This 120 GHz MMIC shows exceptional performance in terms of noise figure (NF) and power consumption. The low-noise

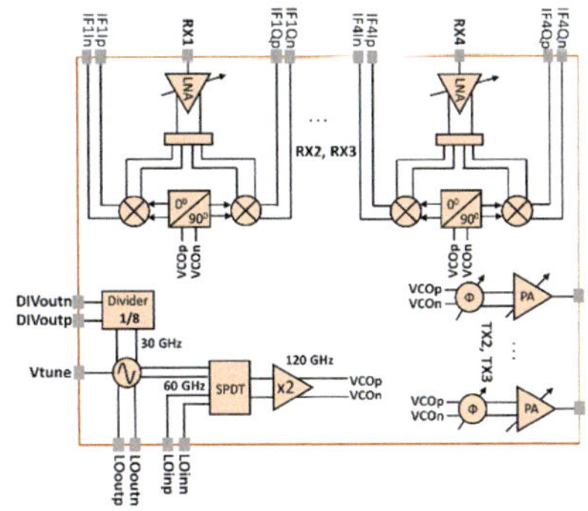

Fig. 6. The block diagram of the designed TRX for the in-cabin radar demonstrator developed in the BEYOND5 project.

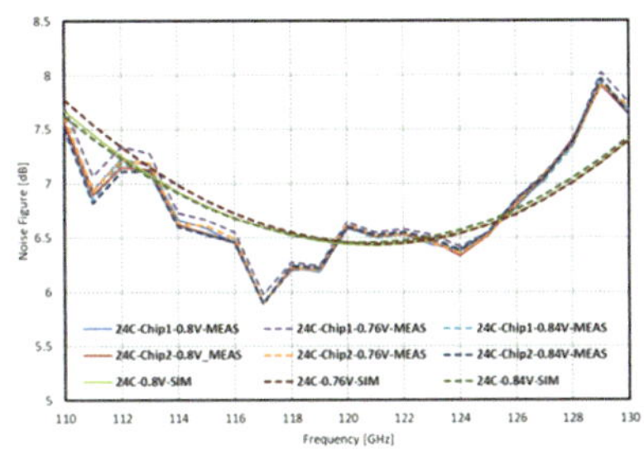

Fig. 7. TRX low-noise amplifier noise performance for the in-cabin radar demonstrator developed in the BEYOND5 project.

amplifier (LNA) achieves a NF of 6.5 dB and a gain of 14 dB in the relevant frequency band, illustrated in Figure 7.

The MIMO 4 × 4 chip consumes a total of 260 mW during continuous operation, accommodating four receivers and one transmitter. The detailed performance metrics are given in Table 1.

III. SMART CUT ON III-V MATERIALS: INNOVATIVE SUBSTRATES FOR THz ERA

The Smart Cut process is also compatible with materials beyond silicon, reducing the reliance on rare and costly bulk materials, thus offering solutions for technological sovereignty aligned with the

TABLE 1.
THE MEASURED MIMO 4 × 4 CHIP PERFORMANCE.

Parameter	120 GHz MPW2
Frequency range	116–123 GHz
TX output power	3 dBm
Conversion gain RX	18 dB
RX noise figure	SSB: 9.1 dB
DC power consumption	260 mW (TX- 36 mW)

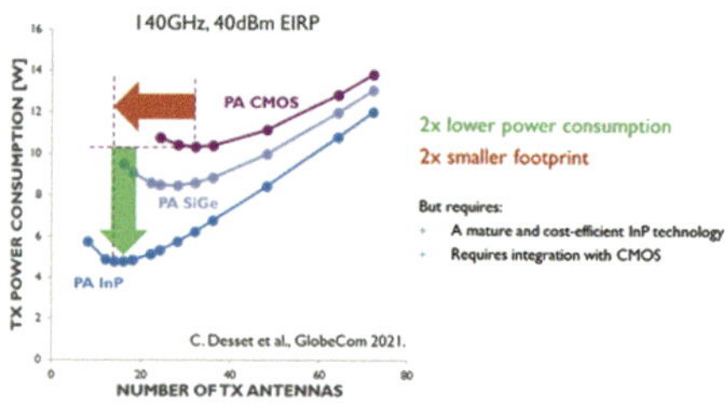

Fig. 8. Power consumption and footprint PA comparison across different technologies at 140 GHz operation frequency and 40 dBm effective isotropic radiated power.

Greendeal initiative. Indium phosphide (InP) offers exceptional capabilities that surpass other technologies in terms of performance and power consumption. It features superior power amplifier (PA) output power (P_{out}) and power added efficiency, along with low NF characteristic. Figure 8 compares PA performance and area metrics of competing technologies like CMOS, SiGe, and InP.

Despite its advantages, the InP technology faces limitations in reaching high-volume production markets primarily due to its manufacturing constraints, small diameters, and high costs. The Smart Cut process provides a sustainable and commercially viable solution for consumer applications like 6G mobile communication and high-resolution sensors. The Smart Cut process in combination with InP paves the way to increase the number of components from the same

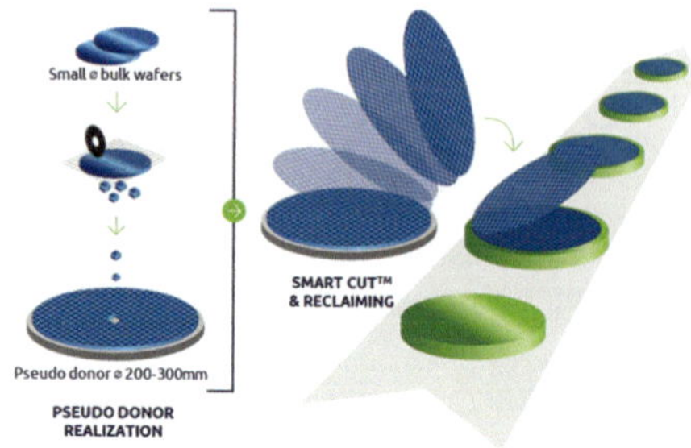

Fig. 9. Schematic of the Smart Cut with tiling approach.

InP wafer by more than tenfold and to transition to larger wafer diameters using a tiling approach.

The Smart Cut with tiling approach is illustrated in Figure 9. This approach involves arranging transferred layers on a larger substrate to optimize material usage and enhance production efficiency. On the left of Figure 9, the production of the pseudo-donor substrate through tiling is illustrated. On the right, the InPoSi substrate is shown, which is obtained from the pseudo-donor substrate via the Smart Cut process, demonstrating the concept of reclaiming the pseudo-donor substrate for multiple reuses.

The “Move2THz” KDT project, initiated on June 1, 2024, unites a European value chain of 28 partners specializing in InP technology development and application, as illustrated in Figure 10.

The Move2THz project aims to revolutionize the manufacturing process of InP platforms by establishing a groundbreaking global standard for InP on silicon (InPoSi). This initiative facilitates the

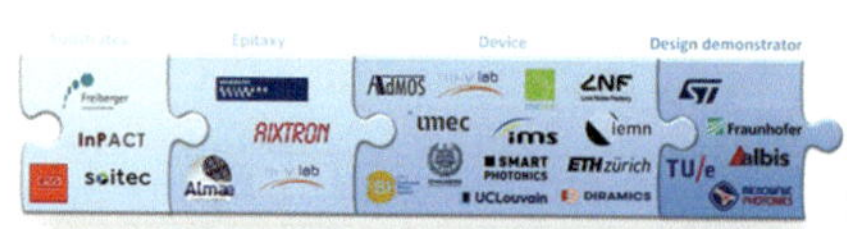

Fig. 10. The Move2THz project consortium illustrated across the value chain.

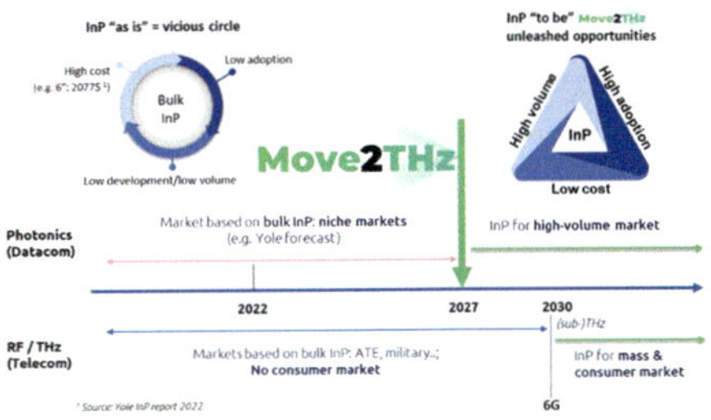

Fig. 11. The Move2THz project aims to break the current vicious cycle, enabling the adoption of InP in large-volume mass markets.

scale-up of wafer size and volume while ensuring compatibility with CMOS standards and minimizing the use of rare InP resources. By establishing a robust value chain from materials to demonstration, Move2THz aims to transform InP from a niche technology into a sustainable and commercially viable platform. This advancement will enable mass-market applications like 6G mobile communication, photonics datacom, RF/bio sensing to utilize frequencies up to THz levels. The methodology on how the Move2THz project aims to enable the adoption of InP in large-volume mass markets is illustrated in Figure 11.

The Move2THz project develops innovative PAs, LNAs, and heterogeneous integration blocks using advanced and environmentally friendly substrates and fabrication processes. These technologies will be integrated and tested for D-Band circuits, sensing, Satcom, non-terrestrial networks, as well as photonics applications.

Alternatively, to InP, RF gallium nitride (GaN) technology also demonstrates significant advantages for RF PAs in 5G base stations using GaN on silicon carbide (SiC) substrates. Soitec has developed a pioneering concept with the Smart Cut process, introducing GaN-on-X substrates that overcome the physical limitations of heteroepitaxial growth. This method involves transferring a thin GaN film onto

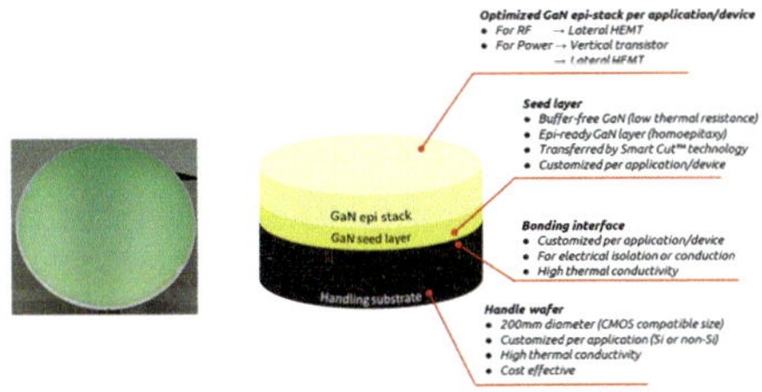

Fig. 12. The SmartGaN substrate stack-up for optimized power or RF applications.

a handling substrate (such as Si, SOI, polySiC, etc.), specifically tailored for RF applications. This innovation enables the creation of devices with superior thermal resistance, higher power density, improved power efficiency, and reduced memory effects with enhancing linearity. The new SmartGaN substrates demonstrated in 8" diameter, address industry concerns regarding GaN quality, CMOS compatibility, thermal conductivity, and cost/availability. This SmartGaN substrate stack-up is illustrated in Figure 12. The value proposal for RF next generation of GaN devices included best-in-class PA performance for 5G and beyond, co-integration of PA and switches, and enabling frequencies beyond 70 up to 100 GHz.

IV. CONCLUSION

European leadership in key markets necessitates the establishment of a complete and reliable supply chain starting from material level up to applications. The industry-proven Smart Cut technology, known for its efficiency in low-power digital applications, sensing and telecommunications, can be adapted to other materials, unlocking application opportunities. By promoting the value chain model, the KDT projects BEYOND5 and Move2THz stimulate synergies and accelerate the development and adoption of new technologies, ensuring a faster time to market.

References

[1] V. Bhagavatula et al., "A 5G FR2 Power-Amplifier With an Integrated Power-Detector for Closed-Loop EIRP Control," in IEEE Journal of Solid-State Circuits, vol. 57, no. 5, pp. 1257–1266, May 2022.

[2] B. Ghyselen et al., "Large-Diameter III–V on Si Substrates by the Smart Cut Process: The 200 mm InP Film on Si Substrate Example," Phys. Status Solidi A, Vol. 219, Issue 4, Feb. 2022.

François Brunier (Member, IEEE) graduated as physics and electronics Engineer from Centrale-Supelec in 1997. From 1998 to 2002, he worked as a device integration engineer for embedded DRAM products in STMicroelectronics Crolles. In 2002, he joined Soitec as head of advanced characterization laboratory. From 2009 to 2011, as a product manager, he led the RF-SOI and power SOI product development and offering. Since 2012, as a partnership program manager, he has been in charge of European collaborative Chips JU programs, IPCEI and public relations.

A CMOS Compatible III-V-on-300 mm Si Technology for Future High-speed Communication Systems: Challenges and Possibilities

A. Vais, A. Kumar, G. Boccardi, S. Yadav, Y. Mols, R. Alcotte, B. Vermeersch, M. Ingels, U. Peralagu, C. Roda Neve, B. Ghyselen, B. Parvais, P. Wambacq, B. Kunert, and N. Collaert

Abstract—**III-V semiconductor-based devices, in particular InP heterojunction bipolar transistors, are a strong contender for next-generation high-speed communication systems. In this paper, we present motivation for the upscaling of III-V technology on to 300 mm Si platforms. A comparison of various options for such III-V on Si technology is described. The challenges in the way to achieve its integration into existing CMOS platform and possibilities to overcome them are shown. We describe imec's path to demonstrate a CMOS compatible III-V-on-300 mm Si technology with the most recent results.**

This work is related to Move2THz, a project supported by the Chips Joint Undertaking (Chips JU) under grant agreement number 101139842. The Chips JU receives support from the European Union's Horizon Europe research and innovation program and the National Authorities.

Corresponding author: A. Vais (e-mail: Abhitosh.Vais@imec.be)

A. Vais, A. Kumar, G. Boccardi, S. Yadav, Y. Mols, R. Alcotte, B. Vermeersch, M. Ingels, U. Peralagu, B. Parvais, P. Wambacq, B. Kunert, and N. Collaert are with IMEC, Leuven, 3000 Belgium.

B. Parvais, P. Wambacq, N. Collaert are also with Vrije Universiteit Brussels, Brussels, Belgium.

C. R. Neve, B. Ghyselen are with SOITEC, Isere, France.

Index Terms—**III-V, HBT, 6G, InP, III-V on Si**

I. INTRODUCTION

THE integration of III-V compound semiconductor-based heterojunction bipolar transistors (HBTs) on Si substrates has been identified as one of the promising routes to enable hybrid III-V/CMOS technology for future RF applications [1–3]. GaAs/InGaP HBTs grown on small size native GaAs substrates are already being used in power amplifiers in mobile phones due to several advantages over other possible material combinations. However, future technologies for high-speed communication applications, such as 6G, would require:

- Better performance from such devices in f_t, f_{max}, P_{out}, and PAE
- Large volume production at significantly lower cost
- Flexibility in circuit design
- Smaller chip footprint [1,2].

All these requirements can be achieved by the monolithic growth of III-V materials on the Si substrate and its integration with state-of-the-art CMOS back-end-of-the-line process for routing complex control signals. As far as III-V growth on Si substrate is concerned, several techniques have been explored; examples are strain relaxed buffer (SRB) layers on Si wafers [4], GaAs grown on Si/SiGe-based compositionally graded buffers [5], confined epitaxial lateral overgrowth (CELO) [6], and aspect ratio trapping (ART) inside the narrow trenches [7]. A review of various monolithic integration approaches is given in [8]. In particular, for RF applications, silicon on lattice engineered substrates (SOLES) [9] have been used to demonstrate InP HBTs co-integrated with CMOS but has the disadvantage of requiring expensive starting substrates and very thick buffer layers to grade to the InP lattice constant, which contradicts with cost and complexity requirements of such future technologies. It is imperative, then, that in order to minimize the complexity and cost of a hybrid CMOS/III-V technology, standard Si substrates should be used.

Furthermore, with ever-growing requirement for larger bandwidths for mm-wave applications, InP-based HBTs are gaining importance [10]. It is then obvious that a cost-effective hybrid CMOS/III-V technology on a standard 300 mm Si substrate, which can be extended to InP-based material system, would be highly desirable.

In this work, we present a comparison of various options for a hybrid III-V/CMOS on Si technology. Various challenges in the way to achieve such an integration into existing CMOS platform and possibilities to overcome them are shown. Furthermore, we report imec's path to demonstrate a CMOS compatible III-V-on-300 mm Si technology consolidated with the most recent electrical characterization results of InP HBTs.

II. III-V ON SI SUBSTRATES: OPTIONS AND CHALLENGES

One of the major challenges related to the integration of III-V materials on Si substrates is the large mismatch in lattice constant as compared to Si (4% for GaAs and 8% for InP and InGaAs) and thermal expansion coefficient, which creates a large number of relaxation defects in the semiconductor layers [8]. These defects have been shown to severely limit the electrical performance of devices [6]. This has led the research on the techniques to integrate III-V materials on Si to diversify into two broad categories: heterogeneous integration, where the challenge is circumvented by directly bonding a high quality III-V layer or substrate to the Si substrate, and monolithic integration, where III-V material is directly grown on Si. In both categories, there are several integration options among which, in the following section, the most promising techniques are described in more detail.

A. Heterogeneous Integration

The heterogeneous integration techniques, an example of which is shown in Figure 1, involve growing a III-V device stack on native III-V (GaAs or InP) substrate, cleaving the device wafers into square tiles, and transferring these tiles by the pick-and-place technique directly onto the large-diameter target Si wafer by a bonding process [12,13]. The sequence of stack growth, cleaving, and bonding might change depending on the technique. The bonding process can be a direct III-V to Si bonding or a dielectric–dielectric layer bonding, such as silicon oxide or silicon nitride. Both techniques suffer from a potential challenge of maintaining highly smooth surface free of small particles. Adhesive bonding involving an adhesive material, such as epoxy,

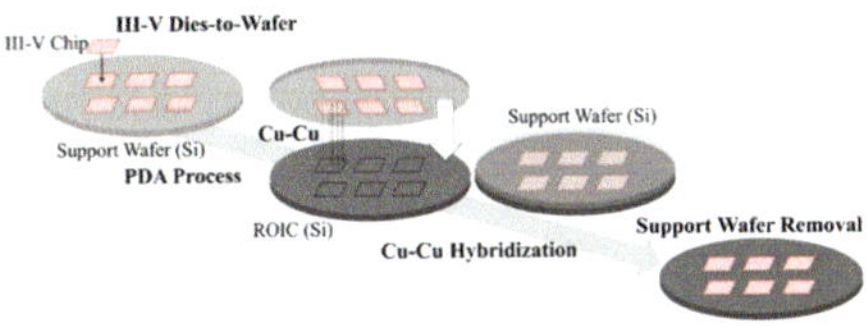

Fig. 1. An example of heterogenous integration for visible-SWIR image sensor with read-out integrated circuit (ROIC). III-V chip can either be just a diced native substrate or a die with already fabricated device [13]. Bonding process can consist of Cu–Cu or oxide–oxide hybrid bonding.

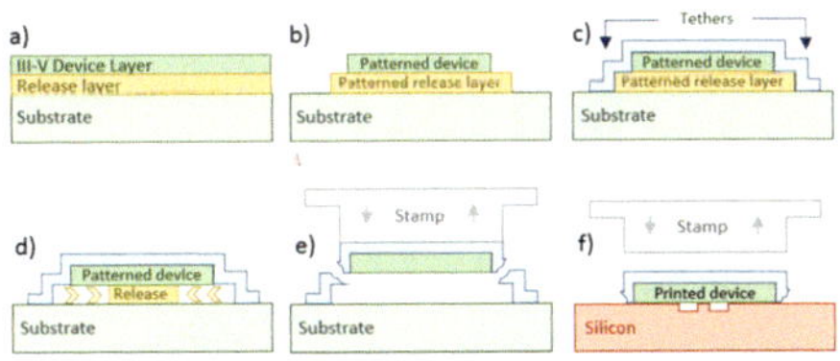

Fig. 2. The μ-transfer printing process. (a) and (b) The definition of the III-V device on the III-V source wafer. (c) and (d) The structures are encapsulated and the release layer is selectively etched. (e) The use of a stamp to pick-up arrays of III-V devices from the source wafer. (f) Print them onto the silicon photonic target wafer, after which the encapsulation is removed and the III-V devices are electrically contacted on a wafer-scale [16].

polyimide, or benzo-cyclobutene has also been used for this purpose, with the advantage of not requiring strict surface cleanliness or smoothness requirements. Another challenge in using this technique is to maintain the stability of the tile/wafer interface during the removal of the InP substrate or subsequent process steps, as the grinding/wet etching process may cause stress and lead to cracks in the III-V device layer. Corners of the tiles are particularly vulnerable during this process step. In the case of adhesive bonding, the temperature range of the bonding material limits the thermal budget of all subsequent process steps. Nevertheless, some of these approaches do show promise in circumventing some of the above issues.

Micro-transfer printing is one such heterogeneous integration approach for transferring partly or fully processed III-V devices to Si-based wafers (Figure 2) [14–16]. It requires the growth of a modified device stack on a release layer; for InP, for example, the release layer may consist of InAlAs. Before the transfer, the InP substrates undergo several process steps following the epitaxial growth of the HBT device stack. In these process steps, device mesas are created and the release layer is selectively removed by wet etch resulting in a device layer attached to the substrate only by polymer tether structures. Using a polymer stamp, the devices are then picked by breaking the tether structures and transferred to the target wafer. This technique is limited by the possible lateral material removal of the release layer. Furthermore, there are concerns regarding the possibility of yield issues due to fragments of the broken tethers and process instabilities. The choice of the adhesive bonding layer is also not straightforward to ensure a stable bonding interface during device processing. Although the accuracy of the printing tools has shown improvements recently, it is still challenging to minimize particle contamination during multiple pick-and-place cycles.

Another promising heterogeneous integration approach is to transfer only a thin InP film to the target Si wafer. This InP film is then used as a starting surface for the epitaxial growth of the device stack. [17]. Such engineered substrates are also referred as InP on silicon (InPOSi). One of the main challenges in such a transfer is related to the release of this thin InP layer from native substrate. One approach is the bonding of a bulk InP substrate, including a top thin InP and etch-stop sacrificial layer, on a Si wafer, followed by a backside wet etching of the full InP bulk material [17]. In such a case, not

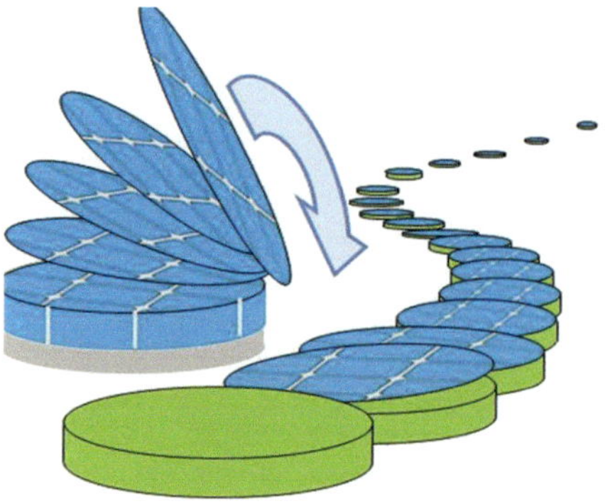

Fig. 3. Simplified sketch of the production of tiled 200/300 mm InPOSi wafers [18].

only the small size of the native InP substrates (4") poses a problem, but the recycling out of the etchant waste, which also includes precious InP material, may also be challenging.

A more cost-effective approach for transferring InP films is described by Ghyselen et al. [18] based on the Smart Cut™ technology initially introduced by Soitec for the fabrication of SOI wafers (Figure 3). In such a process, an InP donor wafer, a Si target wafer, or both are covered with an oxide layer. The InP substrate is then subjected to hydrogen ion implantation, forming a buried layer of microcavities at a defined depth. After the wafer-to-wafer bonding, a thermal treatment is applied to split off the InP substrate at the buried layer and to strengthen the covalent bonds along the bonding interface of the achieved InPOSi wafer. This process leads to a thin InP film on top of the target oxidized Si wafer while circumventing the issues with wet-etch-based substrate removal and saving InP material from going to waste. Although it is still challenging to upscale this technique to wafer sizes of 200–300 mm, there have been some developments in this regard where a 200mm InP-tiles-based pseudo-donor wafer is first created using traditional transfer approaches, which is then used as a native substrate for transfer as described above [18]. Such native pseudo-donor wafers can be used for several cycles of transfer improving the reusability of the donor substrate. A challenge using this technique is related to post-bonding epitaxial growth temperature. As the wafer-to-wafer bonding process is performed at room temperature, heating up to the growth temperature (typically > 550 °C) induces compressive strain in the InP film because of the larger thermal expansion coefficient of III-V materials compared to the one of Si. Consequently, innovative strain management techniques are required for the III-V growth on InPOSi wafers to prevent the formation of misfit defects inside the HBT device stack.

While all the above techniques show some promise in large-scale integration, the cost of InP substrates to be required is a very vital factor to consider in mass production. While a larger market for InP substrates may lead to lower substrate costs, scarcity may result in price hikes. In this respect, it may be interesting to recycle waste products generated during InP grinding, for instance. The sustainability and conversion efficiency of such recycling is a major aspect that should not be ignored in the assessment of upscaling. The reduction of the environmental impact should be a strong driving force in the development of future technologies and requires dedicated life-cycle assessments.

B. *Monolithic Integration*

Given the above-mentioned constraints regarding material cost and sustainability, the most cost-effective approach for integrating III-V materials with Si substrates seems to be the direct monolithic deposition, eliminating the need for InP substrates. However, the mismatch in

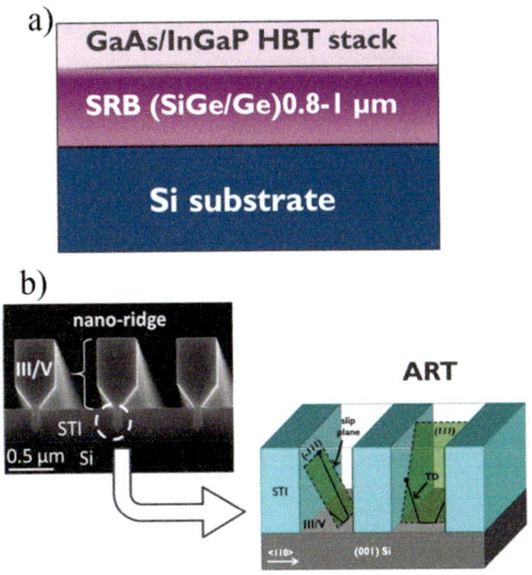

Fig. 4. (a) An example of SRB-based growth of III-V on Si substrate using a thick buffer layer. (b) Cross-sectional SEM image of an NR array. (b) Aspect ratio trapping (ART), where defects are trapped in narrow trenches [19].

lattice constants and thermal expansion coefficients between Si and III-V materials results in the formation of relaxation defects, which adversely affect the performance of HBT devices. The growth of SRB prior to the deposition of the device layers allows to reduce the defect density in HBTs (Figure 4a). In this context, various strategies have been developed to eliminate relaxation defects, including the use of strained layer superlattices as a defect filter, the application of composition gradients, and the use of cyclic annealing treatments. Nevertheless, a low defect density at the buffer surface requires the growth of a several μm thick SRB, which can cause new problems with wafer warpage and crack formation in the III-V layer (9).

Selective area growth (SAG) is an alternative monolithic integration approach that circumvents the issues related to blanket layer growth [6,7]. This method, CELO as an example, involves patterning an oxide layer on top of a Si wafer to expose the Si surface through confined openings. The III-V material is then epitaxially deposited inside these openings under growth conditions that promote selective growth. A high aspect ratio (depth divided by the width) of the patterned openings helps to reduce defects efficiently within a thin layer deposition, as threading dislocations and planar defects (PDs) tend to be trapped at the sidewalls [7]. However, an effective ART of defects inside the openings is accompanied by a small III-V volume confined by the pattern dimension, which poses challenges in realizing an HBT for useful power amplification (9).

Nano-ridge (NR) engineering is a novel integration technique that provides a solution for increasing the volume of III-V material (Figure 4b) [9,19]. It builds upon SAG and ART and involves the application of Si wafers with a narrow trench pattern, where each trench bottom is formed by two {111} facets creating a V-shape, which impedes the formation of anti-phase disorder in the III-V material. A high aspect ratio (>3) of the trenches ensures efficient defect trapping near the III-V/Si interface. The growth is performed by metal-organic vapor phase epitaxy (MOVPE). Once the trenches are filled with III-V material, growth continues out of the trenches to create a large NR above the oxide mask with an increased volume of III-V material. Manipulating the growth rate hierarchy on the different NR facets through adjusting the MOVPE conditions offers the possibility to engineer the shape of the NR. The advantages of using NRE for scaling up the InP technology are evident due to the monolithic III-V integration to Si, which removes the need for InP substrates, thereby reducing cost and potentially enhancing sustainability. The III-V device deposition can be done directly on large-diameter substrates in a 200/300 mm CMOS foundry, without the need of any complicated III-V layer transfer or bonding processes, resulting in a lower risk of particle contamination and yield loss. The application of SAG on patterned wafers makes the growth of thick SRBs unnecessary and enables advanced device scaling and circuit design since the mask layout determines the III-V device distribution throughout the patterned Si wafer. This approach is detailed in the next section.

III. III-V-ON-300 MM SI SUBSTRATE AT IMEC

A. Monolithic Integration using Nano-Ridge Engineering

To integrate III-V materials on 200/300 mm Si substrates, imec is researching on both heterogeneous and monolithic approaches. In the monolithic approach, imec has developed a fabrication process based on NRE, which selectively grows the III-V material in pre-patterned structures or trenches in the Si (see Figure 5).

These high-aspect-ratio trenches are very effective in trapping the defects in the narrow bottom part and growing high-quality, low defect material out of the trench. In addition to a high-aspect-ratio trench to capture defects, a second silicon oxide pattern is added to the original trench pattern, which has the same pitch but a wider opening. After filling the first oxide trench on top of the Si substrate, the MOVPE parameters are adjusted to promote lateral broadening of the NR until it reaches the sidewalls of the second oxide pattern. In this way, the collector material of the HBT is in contact with the oxide sidewall, and subsequent p- and n-doped device layers only grow on the (001) surface without any sidewall deposition. The related illustration is shown in Figure 5c. The successful implementation of this integration concept for a GaAs-based

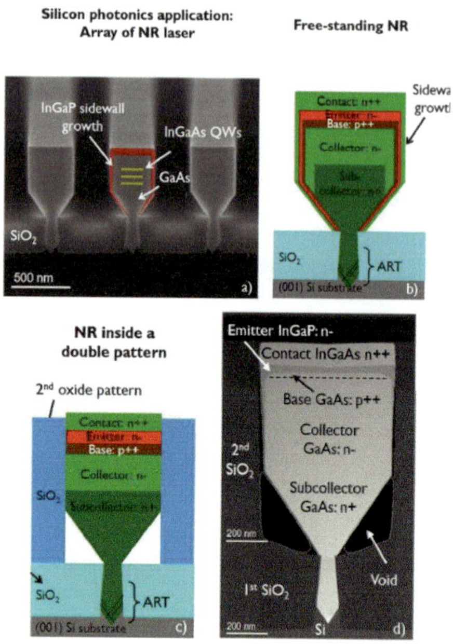

Fig. 5. (a) Cross-sectional SEM image of an NR laser array. (b) The impact of sidewall deposition when an HBT device stack is deposited on a free-standing NR. (c) A second oxide pattern on the first prevents sidewall growth. (d) HAADF-STEM (high-angle annular dark-field scanning transmission electron microscopy) image of a GaAs-based NR HBT [19].

HBT structure grown on a double patterned 300 mm Si wafer is shown in Figure 5d. The threading dislocation density (TDD) at the surface of NRs deposited in narrow trenches has been reported for GaAs NRs to be below 3 × 106 cm–2 based on electron channeling contrast imaging (ECCI) [19]. This defect density value is limited by the ECCI inspection area; therefore, the actual TDD could be even lower for the GaAs NRs. In the TEM investigation of the NR HBT device stack, no indication of any misfit or threading dislocations was found outside the STI trench. All mismatch-induced defects were restricted inside the trench and confined close to the GaAs/Si interface by ART. The complete III-V device stack in the second oxide is free of TDs. Only PDs

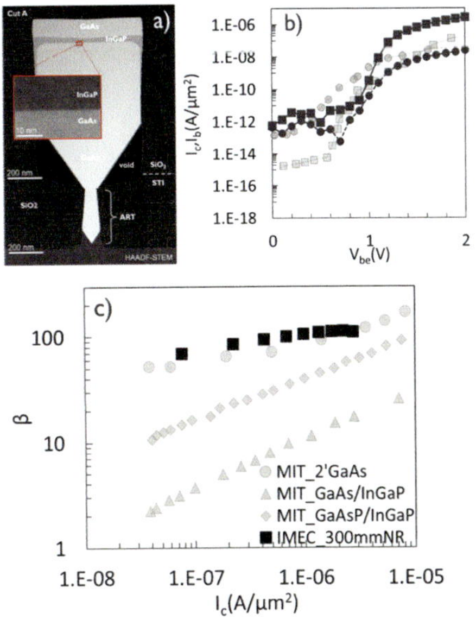

Fig. 6. (a) HAADF-STEM of GaAs/InGaP HBT stack after epitaxial deposition on 300 mm (100) silicon substrate along and across a nano-ridge. (b) A comparison of electrical characterization (Gummel plot) of the devices fabricated in this work with that of ref. [6]. (c) A comparison of DC current gain, β, of our HBT with that of ref. [6] with respect to collector current [20].

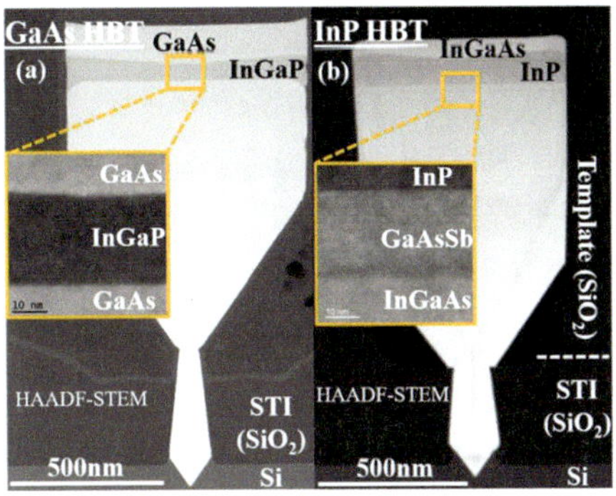

Fig. 7. HAADF-STEM pictures of (a) GaAs and (b) InP HBT stacks on 300 mm Si substrate across a single nano-ridge (NR) showing good device layers interface quality [23].

such as stacking faults and micro-twins could be found in the device stacks laying in a {111} plane perpendicular to the trench. The density of 0.5–0.9 μm^{-1} (PD per NR length) is slightly higher as it was observed for GaAs NRs deposited on single oxide pattern [19]. However, as PDs do not involve either open crystal bonds or a pronounced strain field, we expect less impact on the device performance in comparison to dislocation defects.

Initially, HBT devices on such stacks were demonstrated to be fabricated on small coupons taken from these 300 mm wafers (Figure 6) [20–22]. The ideality factors of NRE-based emitter-base and base-collector diodes were ~1.2 and 1.4, respectively, indicating the good quality of EB and BC junctions. An HBT device showed a peak DC gain, β, of 112 at V_{be} = 2 V. A breakdown voltage, BV_{cbo}, of 10 V was also measured on this device. The quality of the layer interfaces, verified using electrical characterization, was found to be similar to the same stack grown on high-quality GaAs substrates. Furthermore, HBTs fabricated on this stack showed an electrical performance considerably better than GaAs(P) devices fabricated on a Si substrate with SRB layers, without any need to grow thick (>1–10 μm) buffer layers.

Recently, imec demonstrated the first monolithic integration of III-V HBTs on 300 mm Si substrate using standard fabrication in a CMOS pilot line for both GaAs and InP HBTs (Figure 7) [23]. The fabrication for both material systems was done completely in a 300 mm CMOS pilot line at imec with excellent electrical characterization results with low across-the-wafer and wafer-to-wafer non-uniformity (Figure 8). The ideality factors of collector and base current for GaAs HBTs are ~1.0 and 1.45, respectively (as compared to ~1.02 and ~1.6 on GaAs substrate), indicating on the effectiveness of the integration in maintaining the good quality of hetero interfaces. A good DC current gain, β_{GaAs} ~ 40 and β_{InP} ~ 7, and ON

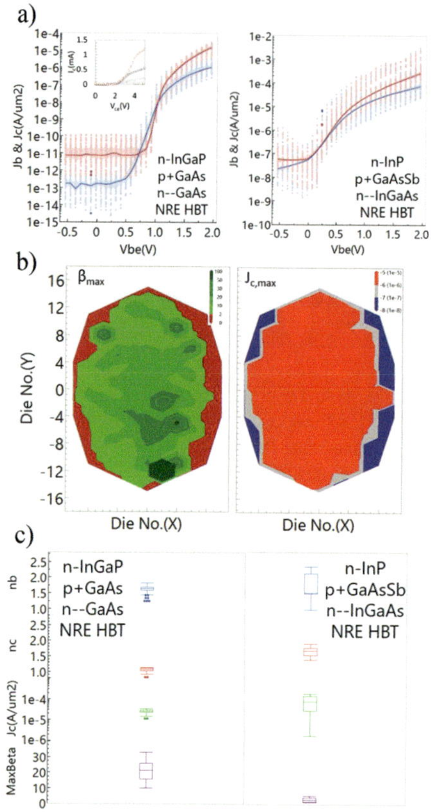

Fig. 8. (a) Gummel plots of HBTs fabricated on GaAs and InP NRE material stack. The output characteristic of GaAs HBT is also shown in the inset. (b) Across wafer variation of DC current gain and maximum collector current of a GaAs NRE HBT wafer. (c) Key DC parameters variation of GaAs (left) and InP (right) NRE HBTs [23].

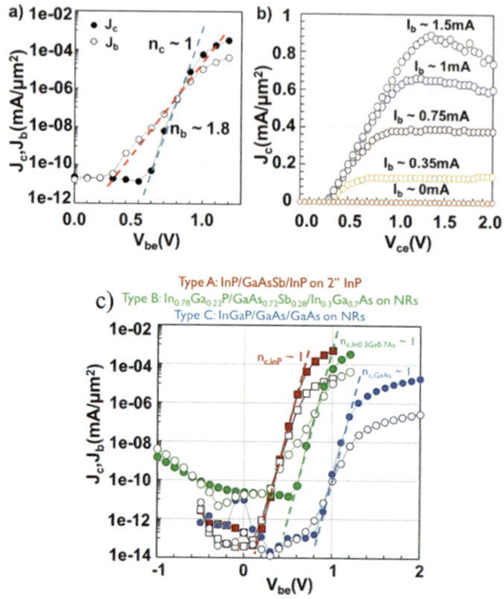

Fig. 9. (a) Transfer and (b) output characteristics of an NR-HBT fabricated on $In_{0.78}Ga_{0.22}P$/ $GaAs_{0.72}Sb_{0.28}$/ $In_{0.30}Ga_{0.70}As$ stack. (c) A comparison of the Gummel plots of HBTs fabricated on three different III-V material systems to illustrate the improvement of on current in $In_{0.3}Ga_{0.7}As$ NR HBTs as compared to GaAs NR HBTs [24].

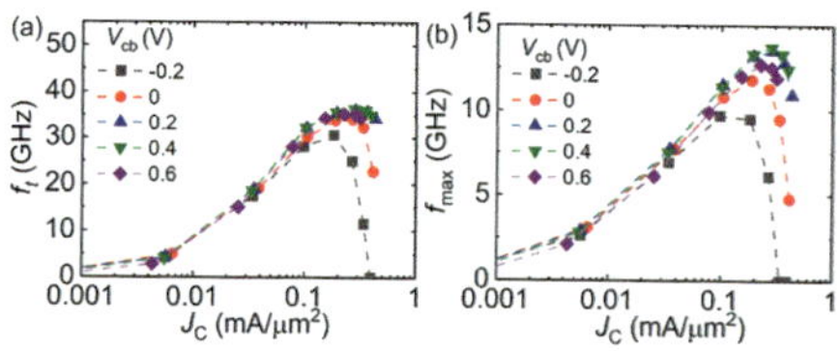

Fig. 10. (a) f_t and (b) f_{max} for a device with W_E: 5 µm and 55 nano-ridges (L_E: 44 µm) as functions of current density J_C. At V_{cb} = 0.4 V, extracted peak f_t is ~36 GHz and peak f_{max} is 13.7 GHz [24].

current density, $J_{c,GaAs}$ ~ 6 × 10^{-5} A/µm^2, $J_{c,InP}$ ~ 3 × 10^{-4} A/µm^2 is observed for GaAs and InP devices, respectively. Moreover, a BV_{CEO} and BV_{CBO} of ~8 and ~13 V, respectively, are measured on GaAs NRE HBTs. These values, although not state-of-the-art, are comparable to similar devices made on 2" native substrates. By demonstrating it for two different III-V material systems, GaAs and InP, imec showed this technology's flexibility to cover several III-V based application domains ranging from FR3, 6G, and optical I/O systems.

In addition, imec also demonstrated, recently, an example of this versatility by fabricating an NRE-based Ga-rich $In_{0.78}Ga_{0.22}P/GaAs_{0.72}Sb_{0.28}/In_{0.3}Ga_{0.7}As$ HBTs on 300 mm Si substrates, which can achieve better performance as compared to InGaP/GaAs HBTs FR-2 and FR-3 5G applications as shown by the DC and RF electrical performance of these devices (Figures 9 and 10) [24]. The ideality factors of collector and base currents are ~1.0 and ~1.8, respectively. The device showed a DC current gain of ~10 at V_{be} = 1.2 V and breakdown voltages, BV_{CEO} and BV_{CBO}, of ~ 2.5 and 4.5 V, respectively. At V_{cb} = 0.4 V, extracted peak f_t is ~36 GHz and peak f_{max} is 13.7 GHz for a device with emitter width W_E: 5 µm and with 55 NRs (L_E: 55 × 0.8 µm = 44 µm).B. Heterogeneous Integration using InPoSi

Due to InPoSi's capability to upscale to large wafer sizes without any issues of release layer removal and its advantages related to sustainability and reusability, imec is also investigating InPoSi's application in developing a III-V on Si technology. In this direction, HBT stack is grown on a 4" InPoSi substrate and on a 4" InP substrate (for comparison) using MOCVD. The InP layer bonded onto the SOI wafer to create the 4" InPoSi wafer was taken from a 4" InP substrate from the same batch as the 4" InP substrate used here for comparison. The RMS surface roughness of the HBT stack on InPoSi, measured using AFM, is observed to be similar to that on InP substrate (~0.13 nm) indicating good epitaxial growth. Large-area HBT devices fabricated on the above stack showed good DC performance. The ideality factors of collector and base currents were ~1.0 and 1.7 (as compared to ~1.0 and 1.3 for the 4" reference InP substrate) respectively indicating the good quality of hetero-interfaces. An HBT device showed a peak DC gain, β, of ~100 is observed at Vbe = 1 V. The breakdown voltages, BVceo and BVcbo, of 3.5 and 5.5 V, respectively, were also measured on this device. A comparison of the DC performance of HBTs on InPoSi and InP substrate is shown in Figures 11 and 12. Further investigation on the RF performance of HBTs on InPoSi is underway. Nevertheless, negligible differences in the performance of HBTs on InPoSi substrate from that on native substrate points to promising future of this technique in the development of III-V on Si technology.

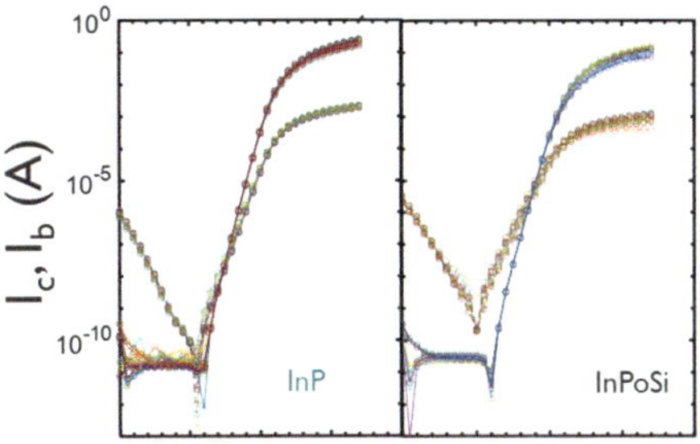

Fig. 11. Transfer characteristics of a InP/GaAsSb/InP HBT fabricated on InP (left) and InPoSi (right) substrates. Except for the difference in base current ideality factor and leakage, ON state performances of both are quite similar and good.

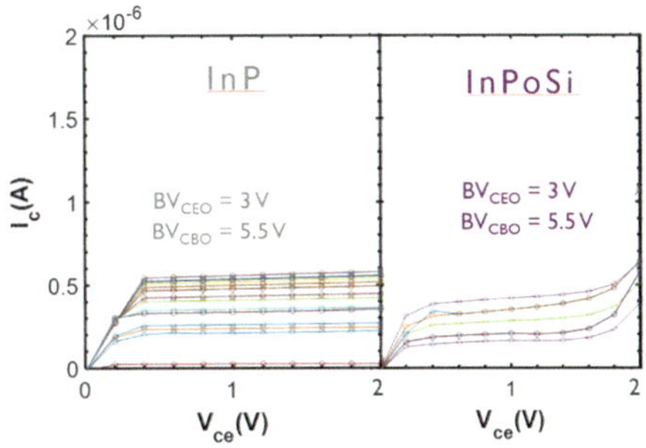

Fig. 12. Output characteristics of an InP/GaAsSb/InP HBT fabricated on InP (left) and InPoSi (right) substrates for the same base currents. Note the slight difference in the output current.

V. CONCLUSION

The cost-efficient and sustainable integration of III-V on Si is critical for the success of future high-speed communication systems. Different approaches have been reviewed, with a special emphasis on imec's recent developments, which are at the forefront of the research to achieve a hybrid III-V/CMOS technology. Compared to the other approaches, NR engineering (for monolithic approach) and InPoSi (for heterogeneous integration) look promising. While CMOS is compatible, both approaches highlighted excellent DC and RF performance of HBTs fabricated on a variety of III-V material systems.

ACKNOWLEDGMENT

The authors gratefully acknowledge the contributions and support of imec members from 300 mm cleanroom, III-V lab and the AMSIMEC test lab. The authors would also like to acknowledge the support of SOITEC under the imec-SOITEC joint development program. SOITEC wishes to acknowledge the teams at SOITEC, SOITEC LAB, and CEA Leti for the development of InPoSi substrates. This work is related to Move2THz, a project supported by the Chips Joint Undertaking (Chips JU) under grant agreement number 101139842. The Chips JU receives support from the European Union's Horizon Europe research and innovation program and the National Authorities.

References

[1] E. Fitzgerald, et al. "Monolithic III-V/Si Integration." ECS Transactions 19.5 (2009): 345.

[2] N. Cahoon, P. Srinivasan, and F. Guarin, "6G Roadmap for Semiconductor Technologies: Challenges and Advances", *IEEE Int. Reliab. Phys. Symp. Proc.*, p.11B1-1–11B1-9 (2022)

[3] N. Collaert, A. Alian, A. Banerjee, V. Chauhan, R. Y. ElKashlan, B. Hsu, M. Ingels, A. Khaled, K. V. Kodandarama, and B. Kunert, "The Revival of Compound Semiconductors and How They Will Change the World in a 5G/6G Era", *ECS Trans.*, 98 (15), p. 15–25 (2020)

[4] T. Oka et al. "Advanced performance of small-scaled InGaP/GaAs HBT's with f_T over 150 GHz and f_{max} over 250 GHz." *IEEE International Electron Devices Meeting 1998. Technical Digest (Cat. No. 98CH36217)*. 1998.

[5] M. L. Huang, , et al. "High performance $In_{0.53}Ga_{0.47}As$ FinFETs fabricated on 300 mm Si substrate." *2016 IEEE Symposium on VLSI Technology*, 2016.

[6] C. Heidelberger, and E. A. Fitzgerald. "GaAsP/InGaP HBTs grown epitaxially on Si substrates: Effect of dislocation density on DC current gain." *Journal of Applied Physics* 123.16 (2018).

[7] L. Czornomaz, et al. "Confined epitaxial lateral overgrowth (CELO): A novel concept for scalable integration of CMOS-compatible InGaAs-on-insulator MOSFETs on large-area Si substrates." *2015 IEEE Symposium on VLSI Technology (VLSI Technology)*. 2015.

[8] N. Waldron, et al. "Gate-all-around InGaAs nanowire FETS with peak transconductance of 2200μS/μm at 50nm Lg using a replacement Fin RMG flow." *2015 IEEE International Electron Devices Meeting (IEDM)*. 2015.

[9] B. Kunert, et al. "How to control defect formation in monolithic III/V hetero-epitaxy on (100) Si? A critical review on current approaches." *Semiconductor Science and Technology* 33.9 (2018): 093002.

[10] J. R. LaRoche, et al. "Monolithically integrated III-V and Si CMOS devices on silicon on lattice engineered substrates (SOLES)." *Proceedings of the International Conference on Compound Semiconductor Manufacturing Technology (CS MANTECH)*. 2009.

[11] M. J. W. Rodwell, et al. "100-340ghz systems: Transistors and applications." *2018 IEEE International Electron Devices Meeting (IEDM)*, 2018.

[12] R. Jones, et al., "Heterogeneously Integrated InP/Silicon Photonics: Fabricating Fully Functional Transceivers", *IEEE Nanotechnol. Mag.*, 13 (2), p. 17–26 (2019)

[13] S. Manda, et al., "High definition Visible-SWIR InGaAs Image Sensor using Cu-Cu Bonding of III-V to Silicon Wafer", *Tech. Dig. - Int. Electron Devices Meet. IEDM*, p. 390–393 (2019)

[14] M. Meitl, et al., "Transfer printing by kinetic control of adhesion to an elastomeric stamp", *Nat. Mater.*, 5 (1), p. 33–38 (2006)

[15] A. D. Carter, et al., "Micro transfer-Printed InGaAs/InP HBTs Utilizing a Vertical Metal Sub-Collector Contact", *Device Res. Conf. - Conf. Dig. DRC*, (6), p. 45–46 (2019)

[16] B. Corbett, R. Loi, J. O'Callaghan, and G. Roelkens, "Transfer Printing for Silicon Photonics", *Semicond. Semimetals*, **99**, p. 43–70 (2018)

[17] C. Besançon, et al.. "Comparison of AlGaInAs-Based Laser Behavior Grown on Hybrid InP-SiO2/Si and InP Substrates", IEEE PTL, vol. 32, no. 8, April 15, 2020.

[18] B. Ghyselen, et al., "Large diameter epi-ready InP on Si (InPOSi) substrates", in *Digests of 2023 International Conference on Compound Semiconductor Manufacturing Technology* (2023)

[19] B. Kunert, et al. "Integration of III/V hetero-structures by selective area growth on Si for nano-and optoelectronics." *Ecs Transactions* 75.8 (2016): 409.

[20] A. Vais, et al., "First demonstration of III-V HBTs on 300 mm Si substrates using nano-ridge engineering", *Tech. Dig. - Int. Electron Devices Meet. IEDM* (2019)

[21] Y. Mols, et al., "Monolithic Integration of Nano-Ridge Engineered InGaP / GaAs HBTs on 300 mm Si Substrate", *Materials*, **14**, 5682, p. 1-16 (2021)

[22] S. Yadav, et al., "DC and RF Characterization of Nano-ridge HBT Technology Integrated on 300 mm Si Substrates", *EuMIC 2020 - 2020 15th Eur. Microw. Integr. Circuits Conf.*, p. 89–92 (2021)

[23] G.Boccardi, et al., "A monolithic III-V on Si integration technology utilizing 300 mm CMOS platform", 2024 *CS Man Tech,* May 2024, p4.2.3

[24] A. Kumar, et al., "An adaptable In(Ga)P/Ga(Sb)As/Ga(In)As HBT technology on 300 mm Si for RF applications", *2024 IEEE MTT-S International Microwave Symposium*, p 323-KH308

Abhitosh Vais received the Ph.D. degree in applied physics from the KU Leuven, Belgium, in 2016, for his work on III-V MOS devices and defect characterization. He joined imec, Belgium, in 2016, where he leads the development of III-V on Si technology for advanced RF applications. Dr. Vais has authored or co-authored more than 60 technical journal papers and conference publications and several patents.

Annie Kumar received the M.S. and Ph.D. degrees in electrical and computer engineering from the National University of Singapore (NUS), Singapore, in 2014 and 2018, respectively. Her work mainly focuses on the integration of III-V transistors on large area silicon (Si) substrates. She is currently working at imec, Belgium, to develop low-cost, high-frequency III-V heterojunction bipolar transistors (HBTs) on large-area Si substrates for 6G applications.

Guillaume Boccardi received the M.Sc. degree in physics and chemistry from the University of Toulon, France, in 2001 and the M.S. degree from the Ecole Nationale Superieure de Physique de Grenoble, France, in 2003. In 2005, he joined Philips Research in Leuven, Belgium, as a research scientist, where he worked on process integration and electrical characterization of HBTs, MOSFETs, HVTs, and MEMS. Since 2011, he has been with imec, Leuven, as a R&D engineer working on MGFETs and currently on III-V devices integration and electrical characterization.

Sachin Yadav has been an RF device researcher at imec, Belgium, since 2019. He is working on design, characterization, and modeling of III-V and III-N devices for RF applications. He received the Ph.D. degree from the National University of Singapore, Singapore, in 2017, where he worked on device design and process integration for III-V optoelectronic devices. As a postdoctoral fellow at the Niels Bohr Institute, Denmark, he worked on the process integration of semiconductor–superconductor hybrids for topological quantum computing applications.

Yves Mols is an epitaxy specialist at imec. After receiving the M.Eng. degree from Universiteit Gent, Belgium, in 2001, he joined imec where he received a doctorate from KU Leuven, Belgium, for his research on the growth of III-V solar cells on germanium. Currently, he is focusing on the deposition of III-V materials on various blanket, patterned, and template substrates for RF applications.

Reynald Alcotte received the Ph.D. degree in semiconductor physics from CNRS, Grenoble, France, in 2018. Then, he joined Surface and Interface Engineered Materials Group, KU Leuven, Belgium, as a post-doctoral researcher from 2018 to 2020 where he worked in the III-V epitaxy on silicon team. In 2020, he joined imec, Belgium, as a researcher where he is currently working on IIIV-on-Si epitaxy. His research interests are on integration of III-V on silicon for optoelectronics and radiofrequency applications.

Bjorn Vermeersch is a principal member of technical staff at imec, Belgium, where he works as a thermal modeling specialist. He received the M.Sc. and Ph.D. degrees in electronic engineering from Ghent University, Belgium, in 2005 and 2009, respectively, and subsequently worked at UC Santa Cruz, Purdue and CEA-Grenoble before joining imec in 2018. His activities focus on non-diffusive heat transport effects in scaled logic and RF technologies, multiscale methods enabling transistor-to-system level modeling, and development of thermal simulation codes that assist in these endeavors.

Marc Ingels (S'88–M'06) received the M.Sc. and Ph.D. degrees in microelectronics from the ESAT-MICAS Laboratories, KU Leuven, Belgium, in 1990 and 2000, respectively. In 1999, he joined Alcatel Microelectronics, which was later acquired by ST-Microelectronics, and was involved in the integration of ADSL analog front ends and Bluetooth RF transceivers. He joined imec, Leuven, in 2005, where he is currently a principal member of technical staff with the Wireless Research Group. He is involved in RF transceivers with a special interest in digital transmitters. He has authored and co-authored over 50 papers. He contributed to a number of book chapters and holds several patents.

Uthayasankaran Peralagu obtained the M.Eng. degree in electronics and electrical engineering and the Ph.D. degree in electronics and nanoscale engineering from the University of Glasgow, Glasgow, UK, in 2008 and 2016, respectively. His Ph.D. research was on the design and development of III-V MOSFET devices for logic applications which he continued to pursue as a postdoctoral associate with the University of Glasgow. Since 2018, he has been with imec, Leuven, Belgium, working on III-N device design and development for RF applications. He is currently an R&D team lead responsible for III-N/III-V device activities in the advanced RF program at imec.

Cesar Roda Neve received the M.Sc. Engineering degree from the ICAI Universidad Pontificia de Comillas, Madrid, Spain, in 2000 and the Ph.D. degree in engineering sciences from the UCL, Belgium, in 2012. From 2004 to 2006, he was with the University Carlos III of Madrid, Spain, where he worked on ROF links and optoelectronic devices. From 2006 to 2012, he was with the Microwave Laboratory at the UCL, Belgium, where he worked on the characterization and applications of HR-Si, HRSOI, and trap-rich HR-SOI substrates. From 2013 to 2016 he was with imec, Leuven, Belgium, focusing on 2.5D, 3D stacking, and packaging. From 2016 to 2021, he worked at M3 Systems Belgium, as a senior project manager on GNSS related projects and satellite/UAV communications. In 2021, he joined SOITEC as R&D Program Manager working on strategic research on RF, 6G, quantum, and advanced CMOS technologies.

Bruno Ghyselen received the Ph.D. degree in materials science in 1992. His work was realized within THOMSON-CSF/ Laboratoire Central de Recherches/Orsay France and was focused on the realization and characterization of high T_c superconducting junctions. He pursued this collaborative work in the University of Cambridge, Cambridge, UK, as a research associate. He joined SOITEC in 1995. Within SOITEC, he has been in charge of different R&D programs and collaborations with different partners, most of them dedicated to the development of advanced SOI substrates for specific applications and other advanced composite substrates.

Bertrand Parvais is a Principal member of technical staff at imec and a professor at VUB Brussels. He received the electrical engineering and Ph.D. degrees from the UCL, Louvain-la-Neuve, Belgium, in 2000 and 2004, respectively. He joined imec as a device engineer, working on the characterization and modeling of transistors in advanced CMOS technologies for analog and RF applications. From 2009 to 2016, he was involved in the design of mixed-mode and millimeter-wave CMOS circuits. He then led the team responsible for transistor compact model and design-technology co-optimization (DTCO). From 2019, he concentrates on RF technologies for front-end modules, including GaN/Si and its sustainability aspects.

Piet Wambacq received the M.Sc. degree in electrical engineering and the Ph.D. degree from KU Leuven, Belgium, in 1986 and 1996, respectively. In 1996, he joined imec, Leuven, where he is currently a Fellow, working on analog/RF/mm-wave IC design in various technologies for wireless applications. Since 2000, he has also been a part-time professor with VUB Brussels, Belgium. He has authored or co-authored six books and more than 350 articles in edited books, international journals, and conferences. He has been a member of the program committee of several solid-state circuit design conferences. He has been a member of the Program Committee since 2012 of ISSCC of which he was the chair in 2023.

Bernardette Kunert received the diploma and Dr. Rer. Nat. degree in physics from the Philipps University of Marburg, Germany, in 2001 and 2005, respectively. From 2006 to 2013, she worked as a project manager for the start-up semiconductor technology company $NAsP_{III/V}$ GmbH, Germany. In 2013, she joined imec, Belgium, as a principal scientist. Since 2023, she has been a scientific director. Bernardette Kunert has a strong background in epitaxial growth and characterization of III-V heterostructures and in (opto)electronic devices such as laser diodes, modulators, detectors, HEMTs, HBTs, MOSFETs, and multi-junction solar cells. In particular, the monolithic integration of III-V devices on silicon or germanium substrates and the invention of new device concepts have always been at the center of her research interests.

Nadine Collaert is a program director and fellow at imec, Belgium. She is currently responsible for the advanced RF program looking at heterogeneous integration of III-V/III-N devices with advanced CMOS to tackle the challenges of next-generation mobile communication. Before that, she was program director of the LOGIC Beyond Si program focused on the research of novel CMOS devices and new material-enabled device and system approaches to increase functionality. She has a Ph.D. in electrical engineering from KU Leuven, and she has more than 400 publications and holds more than 10 patents in the field of device design and process technology. She is a Member of IEEE.

The Impact of Irradiation on DC Characteristics and Low-Frequency Noise of Advanced SiGe:C HBTs

B. Sagnes, A. Adebabay Belie, M. Bouhouche, J. El Beyrouthy, F. Pascal, J. Boch, P. Chevalier, and D. Gloria

Abstract—**SiGe:C heterojunction bipolar transistors (HBTs) have emerged to address the high demand of high-speed applications in telecommunications and more recently in the space domain. But the ramifying effects of radiation on the DC characteristics and low-frequency noise of SiGe HBTs reveal the degradation in performance and reliability of SiGe HBTs and their associated circuits. Consequently, it is important to understand the impacts of total ionizing dose on SiGe HBTs to design radiation-tolerant and viable electronic devices. This presentation investigates the impacts of X-ray irradiations on advanced SiGe:C HBTs, specifically the ones developed in the B55 BiCMOS technology with f_T/f_{max} of 320/370 GHz, supplied by STMicroelectronics in the framework of the European SHIFT project. Different degradation parameters, such as relative base current and base current noise spectral density, are used in this work. Moreover, the impacts of collector doping, geometry, and annealing will be explored in detail.**

Index Terms—**Annealing, base current, collector doping, degradation, low-frequency noise, X-ray irradiation.**

This work was supported by the Chips Joint Undertaking (Chips JU) under grant agreement number 101096256 (project SHIFT). The Chips JU receives support from the European Union's Horizon Europe research and innovation program and the National Authorities.

The University of Montpellier PRESERVE platform has been funded thanks to the financial support of the Region Occitanie and the European Regional Development Fund.

B. Sagnes, A. Adebabay Belie, M. Bouhouche, J. El Beyrouthy, F. Pascal, and J. Boch are with IES, University of Montpellier, CNRS, 34095 Montpellier, France (bruno.sagnes@umontpellier.fr)

P. Chevalier and D. Gloria are with STMicroelectronics, 38926 Crolles France

I. INTRODUCTION

THE demand for high-speed (HS) wireless communication and niche high-frequency applications is continuously growing at present. Transistors such as SiGe: C heterojunction bipolar transistors (HBTs) and III-V HBTs offer high performance for HS applications although the latter requires complex manufacturing steps, which results in expensive cost. Studies have shown SiGe:C HBTs provide huge performance in wide temperatures, ranging from 4.2 to 573 K [1]. Nevertheless, the radiation-rich environment has a negative impact on the reliability and performance of electronic devices, which has been attracting the attention of the radiation society for many years.

In general, the radiation environment induces two effects on devices, namely single event effect and cumulative effects. The former is caused by a single particle as it passes through or near a sensitive node in a circuit, and results in either a

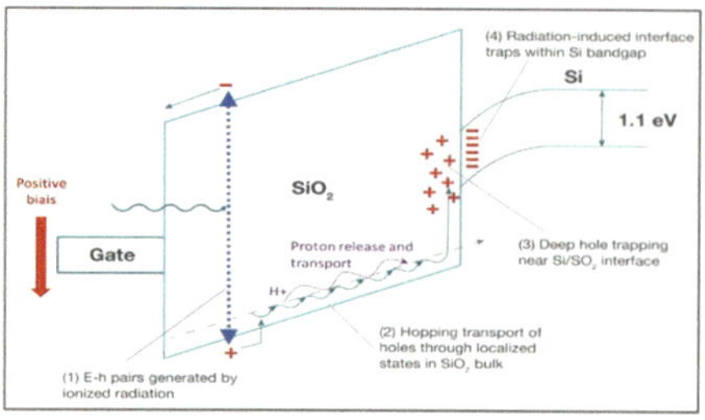

Fig. 1. Process of radiation-induced charge generation in MOS devices [4].

hard error such as single event latch-up or a soft error like single event upsets [2]. Cumulative effects can be divided into two categories: total non-ionizing dose, also known as displacement damage [3], and total ionizing dose (TID), called simply dose, which is the main focus of this study.

The TID effect corresponds to the trapping of charge in the oxide of electronic devices over a period of time leading to the degradation of the component's electrical characteristics. The process that leads to the trapping of charges in semiconductor devices can be illustrated based on a metal–oxide–semiconductor (MOS) structure as depicted in Figure 1.

TID is caused by energetic particle or photon that ionizes the atoms along its path, leading to the generation of electron–hole pairs. Some of these electron–hole pairs recombine due to initial recombination (referred to as charge yield). The uncombined electrons tend to move away from the oxide (in the order of tens of picoseconds) due to their high mobility, whereas holes tend to move slowly to the silicon-oxide interface by hopping mechanism, which results in holes trapping in both oxide and interface traps. These trapped holes in oxide (N_{ot}) and interface traps (N_{it}) later lead to the degradation of the component's electrical characteristics of electronic devices. One of the most important and well-studied effects of N_{ot} is the negative shift in the drain current versus gate-to-source voltage for both N- and P-channel metal–oxide–semiconductor field-effect transistors. The main effect of N_{it} is an increase in the sub-threshold swing of MOS devices.

Unlike MOS transistors, for which the active current layer is at the surface of the semiconductor and in direct link with the Si/SiO$_2$ interface and the oxide layer above it, bipolar devices are "bulk" devices, in that the currents mostly flow through silicon PN junctions. One would therefore expect these bipolar devices to be relatively insensitive to TID, but this is not the case. The most efficient way to highlight TID degradation in a bipolar transistor is to study the influence of increasing TID levels on the Gummel plot, showing the collector and base currents versus the base–emitter voltage, for a constant reverse collector–base voltage. Most of the degradation directly impacts the forward current gain, which decreases with increasing TID level, due to an increase in base current while collector current remains almost unchanged.

As we have seen, trapped charge will have an impact on the behavior of the component, but this effect is not definitive. Indeed, a long time or a high temperature can lead to an annealing effect. Annealing experiments are widely practiced in the radiation community, aiming to remove trapped charges by increasing the temperature of the device over a period of time. Such annealings are also preconized in test methods in accordance with standards [5].

To have a rad-hard device, it is compelling to harden the device either by process or by design as shielding the system from irradiation that leads to an increase in both system weight and complexity, and reduces system reliability [6]. Studies show that SiGe:C HBTs are able to operate in the extreme radiation environment without the need for hardening, with adequate performance and acceptable reliability [7] [8]. Despite this, it is necessary to do extensive characterization in order to model, understand, and investigate the DC characteristics and low-frequency noise (LFNoise) of SiGe HBT in the extreme irradiation environment.

This study presents the impacts of X-ray irradiation on the DC characteristics and LFNoise of SiGe:C HBTs. The impacts of geometry, collector doping, and annealing on the degradation of base current will be explored in the next sections.

II. EXPERIMENTAL SET-UP

The HBTs are issued from a 55 nm CMOS node of BiCMOS technology, B55, developed by STMicroelectronics.

A single-finger CBEBC configuration of HBTs with an emitter area range of 1–4.2 μm² was used for the measurement. The HBTs have three flavors, HS, medium-voltage (MV), and high-voltage (HV), which differ by their selectively implanted collector (SIC) doping, aimed to meet the power and speed requirements [9].

The irradiation experiments were carried out by using 320 keV high-energy X-ray generators located at the PRESERVE platform in the Institute of Electronics and Systems (IES) laboratory, University of Montpellier. It was performed at the dose and dose rate of 40 krad and 11.33 rad/sec respectively in seven steps until it reached a maximum TID of 280 krad. In each step, the forward Gummel measurements and LFNoise experiments were done for approximately one hour. Moreover, the samples were unbiased during irradiation.

After the samples were exposed to a maximum TID of 280 krad, the impact of room temperature was investigated for 10 weeks. After this, annealing experiments were conducted at 100 and 130 °C for a total of 282 annealing hours. To test whether or not thermal annealing has any effect on the DC and LF noise characteristics of non-irradiated transistors, we annealed four pristine transistors at 130 °C for 160 hours. In all cases, no impact was observed.

III. DC CHARACTERISTICS

To understand the static characteristics of the device and determine the range of biasing current or voltage used for LFNoise measurements, it is compelling to first investigate the DC characteristics of the device.

The forward pre-rad and post-rad Gummel characteristics of 0.42 × 10 μm² HBT are presented in Figure 2.

As shown in Figure 2, unlike the collector current (I_C), the base current (I_B) increases with TID. For instance, at 280 krad, I_B is significantly greater than the pre-irradiation I_B as labeled in Figure 2. The significant increase of I_B, subsequently resulting in a reduction of current gain, is majorly shown within the V_{BE} ranging from 0.3 to 0.6 V. While negligible degradation is observed for V_{BE} greater than 0.6 V.

The normalized relative I_B and I_C help to understand the impacts of irradiation quantitatively by comparing the

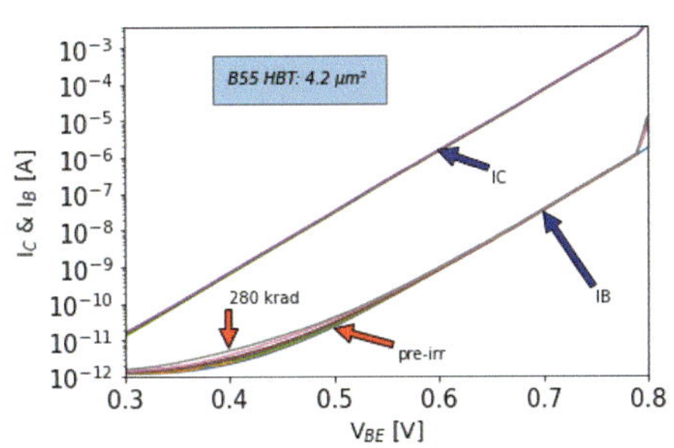

Fig. 2. Pre- and post-irradiation forward Gummel plot of 0.42 × 10 μm² B55 HBT.

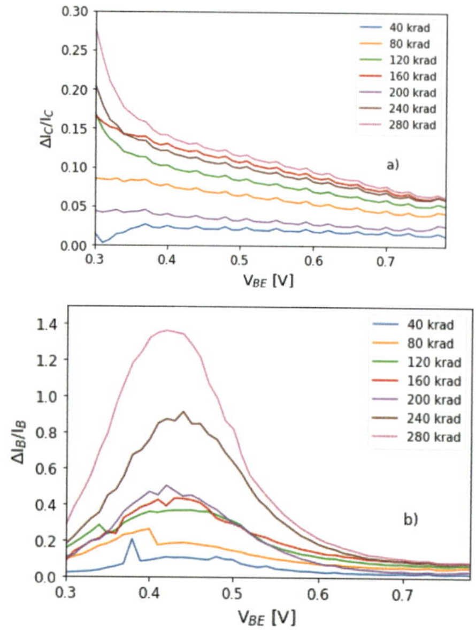

Fig. 3. Examples of: (a) $\Delta I_{C/}I_C$ and (b) $\Delta I_{B/}I_B$ versus V_{BE} at different TIDs.

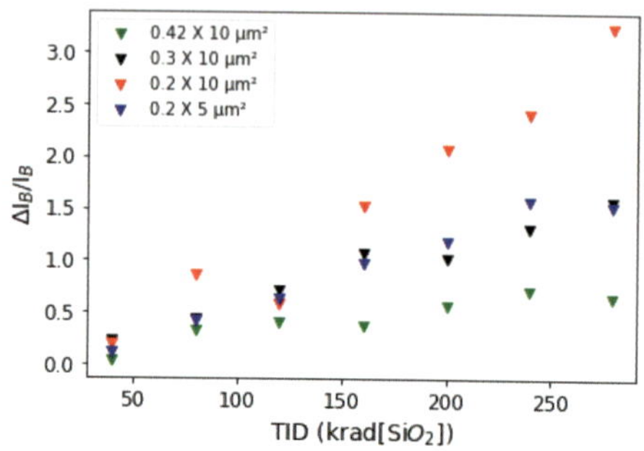

Fig. 4. $\Delta I_{B/}I_B$ versus TID at V_{BE}=0.54 V for HS devices.

post- and pre-irradiation currents, which are expressed as:

$$\frac{\Delta I_C}{I_C} = \frac{IC_{post-irr} - IC_{pre-irr}}{IC_{pre-irr}} \tag{1}$$

$$\frac{\Delta I_B}{I_B} = \frac{IB_{post-irr} - IB_{pre-irr}}{IB_{pre-irr}} \tag{2}$$

The $\Delta I_{C/}I_C$ and $\Delta I_{B/}I_B$ are plotted versus V_{BE} in Figure 3. The normalized collector current, as illustrated in Figure 3a, exhibits a negligible increase with TID. For instance, the maximum degradation was less than 0.1 in the high V_{BE} range. Whereas, the normalized base current showed a maximum increase of 1.4 at a V_{BE} of 0.4 V and TID of 280 krad. But for higher V_{BE}, the degradation of I_B gets very small, especially in the diffusion region where V_{BE} is greater than 0.6 V. Note that for V_{BE} less than 0.6 V, it is the low-injection region where the detrimental impact of irradiation likely happened.

Comparing the normalized collector and base current, it is important to focus on the base current rather than the collector current to investigate the impact of irradiation as the base current shows a significant degradation by irradiation [10].

A. Impact of Geometry

TID has different responses for different geometries of SiGe:C HBTs. Figure 4 is a plot of four geometries HS B55 HBTs $I_{B/}I_B$ versus with TID at V_{BE} of 0.54 V to avoid the influence of tunnel component.

As shown in Figure 4, a linear increase of $\Delta I_{B/}I_B$ with TID is observed. For TID less than 120 krad, the impact of geometry on $\Delta I_{B/}I_B$ is negligible, and this was also shown in previous results [11]. But for TID greater than 120 krad, smaller geometry HBTs showed a greater degradation compared to larger geometry B55 HBTs [12]. Therefore, it could be a better solution to use larger geometry HBTs for rad-hard circuits.

As can be seen in Figure 5, the radiation-induced trap centers in the SiGe:C HBTs potentially could arise in spacer oxides, pedestal oxides, shallow trench isolation (STI), and deep trench isolation. The traps that arise in the emitter–base spacer oxide lead to the depletion

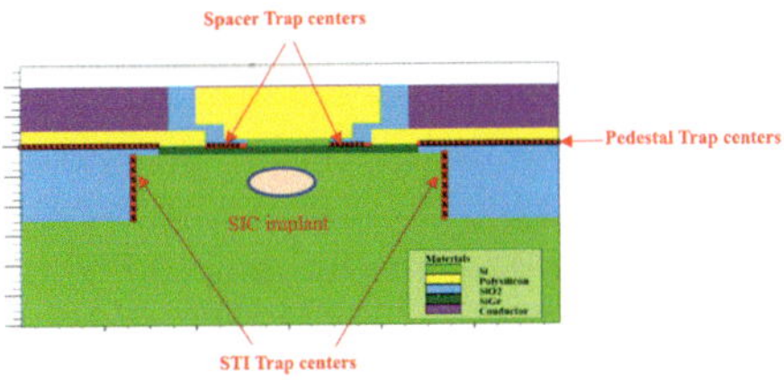

Fig. 5. The possible induced irradiation trap centers in SiGe HBT.

of the emitter–base junctions, which later increase the base current in forward Gummel measurements [13]. Studies have shown that the traps present within STI, and collector–base spacer oxide are responsible for the increase of base current in inverse Gummel measurements [14].

B. Impact of Collector Doping

B55 HBTs are available in three flavors of transistors to meet the power and high-frequency requirements. The transistors are HS, MV, and HV, which differ by their SIC doping.

To investigate the impact of collector doping on the increases of base current of forward Gummel measurements, the $\Delta I_{B/}I_B$ of HS, MV, and HV HBTs are plotted at different TIDs in Figure 6.

As illustrated in Figure 6, the $\Delta I_{B/}I_B$ of HS, MV, and HV HBTs at different TIDs are approximately equal. This in turn leads us to say that the collector doping could have a negligible impact on the degradation of I_B.

C. Annealing

To investigate and study the possible recovery of radiation-induced traps, an annealing experiment was carried out. After the samples were exposed to a maximum TID of 280 krad, the weekly responses had been studied for 10 weeks, to ensure the room temperature has no effect on the recovery of traps. After that, the annealing experiments were performed as summarized in Table 1.

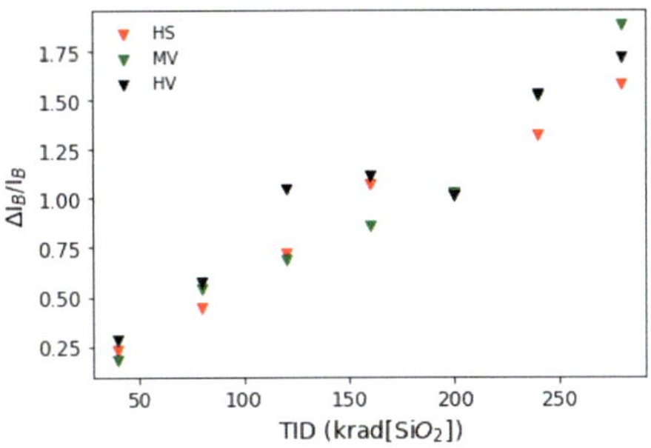

Fig. 6. $\Delta I_{B/}I_B$ versus TID of $0.3 \times 10\ \mu m^2$ HS, MV, and HV HBTs.

TABLE 1
ANNEALING MECHANISM

Annealing temperature (°C)	Annealing hours
100	100
	68
130	24
	90

To understand the degradation and detrapping, $\Delta I_{B/}I_B$ is plotted at three V_{BE} of 0.4, 0.54, and 0.62 V for $0.3 \times 10\ \mu m^2$ HBT.

As shown in Figure 7, the maximum recovery of I_B is approximately 70% at annealing temperature of 100 °C for 168 hours. During the final annealing experiment conducted at 130 °C for a duration of 114 hours, the recovery was noted to be around 90%. Nonetheless, the recovery is different at different V_{BE}. For instance, it is 90% at V_{BE} of 0.4 V, 79% at V_{BE} of 0.54 V, and 55% at V_{BE} of 0.62 V. Note that the degradation of I_B varies with V_{BE} ranges. The highest degradation of I_B was observed at V_{BE} of 0.4 V in comparison to V_{BE} of 0.54 and 0.62 V. This confirms our prior results that the generation–recombination base current component is the main responsible current for the increase of I_B compared to the diffusion base current. But, at high TIDs, the

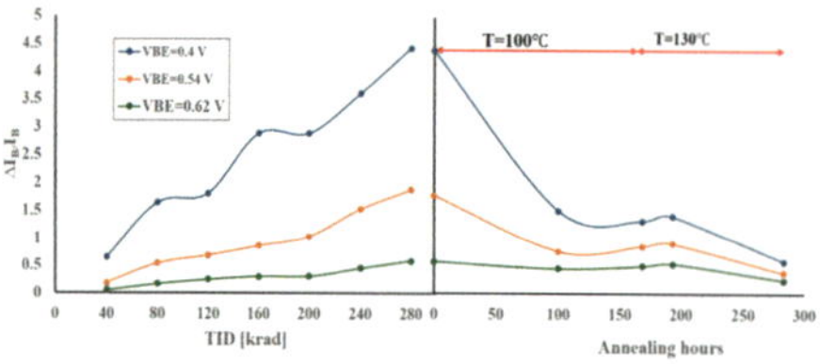

Fig. 7. $\Delta I_B/I_B$ with function of TID and annealing time at $V_{BE} = 0.4$, 0.54, and 0.62 V.

diffusion current also contributes to the degradation of base current. Our measurement result also clearly shows that the room temperature has no impact on the charge detrapping or recovery of radiation-induced traps. For example, in Figure 7, $\Delta I_{B/}I_B$ at zero annealing hour, measured after 10 weeks at room temperature, is approximately equal to $\Delta I_{B/}I_B$ at 280 krad.

IV. IMPACT OF IRRADIATION ON LOW-FREQUENCY NOISE

Compared to electrical *I–V* measurements, LFNoise is a very sensitive characterization technique used to probe and analyze defects, traps, or generation–recombination centers in materials and devices. Hence, it is a very good tool for monitoring the development of a technology and for predicting device reliability [15]. Moreover, due to the nonlinearity in devices, LFNoise can be up-converted and then affects the high-frequency performance of analog circuits such as mixers, oscillators, and amplifiers.

Since the 1990s, intensive LFNoise studies in bipolar junction transistors (mainly PE Si BJTs), III-V heterojunction bipolar transistors (GaAs- or InP-based HBTs), or SiGe HBTs have been undertaken [16] [17]. Concerning the latter, our team has characterized different BiCMOS technologies supplied by STMicroelectronics. In particular, the B55 technology has been extensively studied as part of the European TARANTO project [18]. For part of the current European SHIFT project, we are studying more specifically the effect of X-ray and gamma irradiations on this technology.

First, we present the main LFNoise results obtained on non-irradiated components. Then, we will discuss the results concerning the effect of irradiations, mainly the effect of X-ray TID.

A. Main LFNoise Results Obtained on Pristine B55 HBTs

We directly measured the fluctuations of the input base current, S_{IB}; thus, the expected white noise is the shot noise $2qI_B$. Die-to-die dispersion can lead to different spectra as presented in Figure 8. Excess noise is composed of at least a 1/f component (slope −1), which can be hidden by the presence of one or two generation–recombination components, presented by Lorentzian spectra (plateau + slope −2) and associated with a trapping/de-trapping process [19]. Each spectrum can be analytically approached by the classical deconvolution relation (3), and the second term, relating to the components of G-R, is optional.

$$S_{I_B} = \frac{K_F \cdot I_B^{A_F}}{f} + \sum_{i=1}^{n} \frac{B_i \tau_i}{1 + (2\pi f \tau_i)^2} + 2qI_B \tag{3}$$

Whether as a technological indicator or for circuit simulation, the 1/f noise component has been, and still is, widely studied in comparison with G-R noise. The random presence of the latter makes its study more complicated.

As far as the 1/f noise of HBTs based on B55 technology is concerned, based on our previous studies [20] and on literature data, we can say that:

- the study of K_F versus base bias current and emitter area leads to very

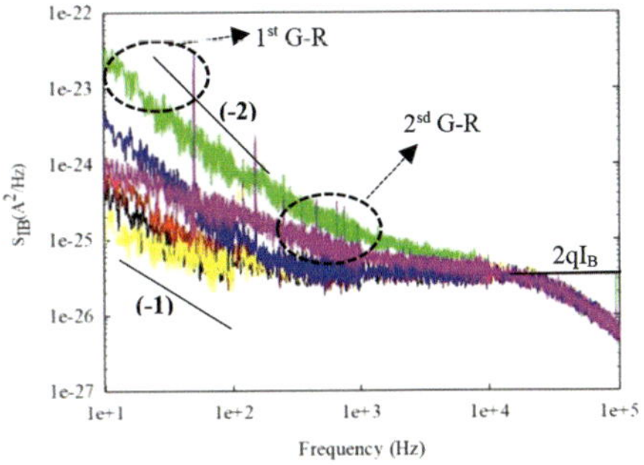

Fig. 8. Base current spectral density for six HBTs with the same geometry of the B55 technology ($A_E = 2\ \mu m^2$ at I_B of 100 nA).

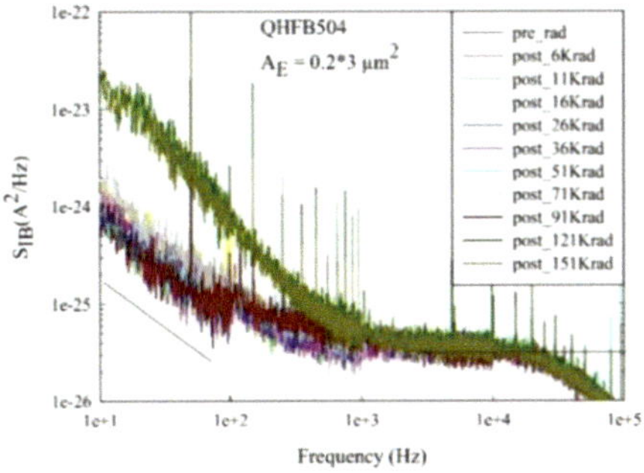

Fig. 9. Various irradiation responses of S_{IB}, for I_B = 100 nA, after X-ray exposure.

good values of the 1/f noise level, with a figure of merit $K_B = K_F \times A_E$ in the range of 2–6 $10^{-10}\ \mu m^2$ close to the best values published [20][21];

- the study of K_F versus different technological parameters (i.e., doping, interface cleaning, etc.) leads to the conclusion that the 1/f noise sources are located at the E-B junction mainly at the poly/mono interface [12][17].

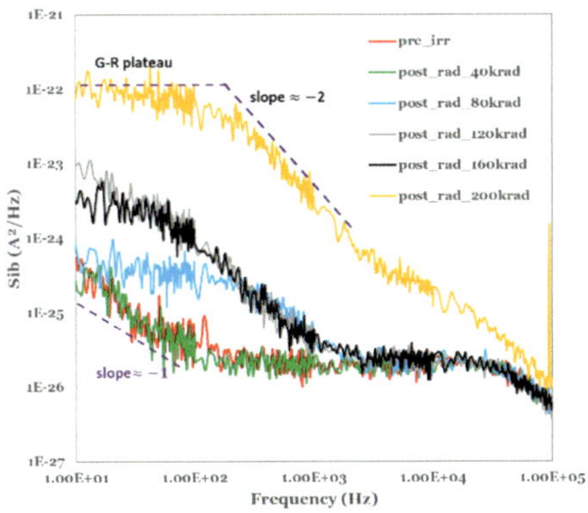

Fig. 10. S_{IB} for I_B = 50 nA at different X-ray TID ($A_E = 3\ \mu m^2$).

B. Post-rad Spectral Analysis

At low TID, below 100 krad and much more rarely up to 150 krad, 1/f noise component is systematically dominated by the presence of one or two G-R components whose cut-off frequency and plateau do not vary much with dose. An example is given in Figure 9. As can be seen from Figure 10, at higher doses (>150 krad) on a few transistors, we observed a very significant increase in the plateau of the second G-R component.

In parallel with the DC characterizations, we studied the effect of thermal annealing on noise spectra after a cumulative dose of 150 krad. Whatever the geometry, we systematically recovered the initial spectra following annealing at 130 °C, notably with the complete disappearance of irradiation-induced G-R components, while native G-R components and 1/f noise were unaffected by annealing.

C. Post-rad 1/f Component

After irradiation, when 1/f noise is unambiguously observed, its amplitude is studied as a function of the emitter surface. As shown in Figure 11, compared with the initial level, the 1/f noise changes very little with dose. Note also that we have always verified the quadratic evolution of noise in 1/f with base current.

We can therefore conclude that sources of 1/f noise, located at the Si/polySi interface, are not affected by X-ray irradiation.

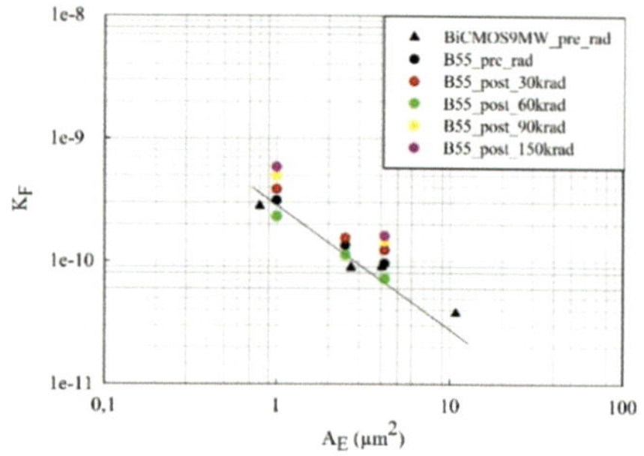

Fig. 11. 1/f noise amplitude, K_F, as a function of A_E for B55 HBTs, before and after irradiation exposure to X-ray. For comparison, the pre-rad of the previous technology (B9MW) is given.

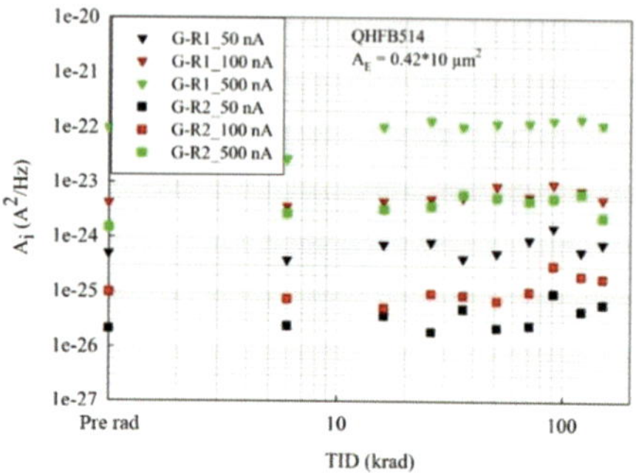

Fig. 12. G-R1 and G-R2 magnitudes as a function of TID.

D. Post-rad Generation–Recombination Components

As can be seen in Figures 8, 9, and 10, when G-R components are present (initial or induced by irradiation), one or two components were found with cut-off frequencies in the range of 20 and 200 Hz. Cut-off frequencies f_{ci} presented an independency to the input current bias as well as to the irradiation dose. Concerning the plateau magnitudes, as can be seen in Figure 12, no significant evolution with the cumulative dose is observed at low dose (i.e., < 150 krad). On the other hand, on the same figure, we can see that the evolution with the base bias current is very pronounced, with an almost quadratic law.

Unlike the 1/f noise, fewer results or physical models exist for initial G-R noise components [22] and, concerning irradiation induced ones, no extended results have been published. Nevertheless, both initial and induced G-R noise sources are suspected at the vicinity of the emitter periphery, E-B spacer oxide, or its interface with the intrinsic E-B junction. Further studies based on technological and geometric parameters are required to confirm this hypothesis.

V. CONCLUSION

The effect of X-ray irradiation on DC characteristics (forward Gummel plot) was observed on the low-injection base current at the recombination–generation component. However, at the highest cumulative dose (i.e., TID > 250 krad), the diffusion regime begins to be impacted (V_{BE} > 0.6 V) and, even if the nominal bias point of the transistor for circuit operation is higher, this could have an influence on its operation and/or reliability. An impact of geometry was observed at high doses (i.e., TID > 120 krad), leading to an increase in base current degradation for the smallest emitter surface areas. In contrast, no significant effect was observed with collector doping. Finally, the various thermal annealing processes recovered more than 80% of the initial characteristics. Following the dynamic of annealing at 130 °C, we can logically expect a 100% recovery after a longer annealing time.

As far as LFNoise is concerned, the effect of X-ray irradiation is reflected in the systematic appearance of one or two recombination–generation components. The level of TID at which G-R components appear fluctuates depending on the transistors, but at a cumulative dose of around 120 krad, all transistors were affected. As a result, 1/f noise was rarely observed above this dose level. Nevertheless, on each occasion, no significant impact on the 1/f noise component was observed. After their appearance, the G-R components showed little sensitivity to low doses of X-ray irradiation. Indeed, we observed that the cut-off frequency as well as the level of these components did not change at all or very little. However, a fairly radical change was highlighted at higher doses (i.e., TID > 160 krad) with, in particular, a strong increase in the plateau of the second G-R. Just as for the Gummel plots, the effect of thermal annealing allowed us to find the initial noise spectra.

Our results, whether in terms of DC characteristics or LF noise, clearly show that the defects induced by X-ray irradiation are localized at the level of the peripheral oxides of the E-B junction. Finally, it should be noted, as we have shown in the reference [11], that gamma irradiations with a cobalt source led to similar results for equivalent cumulative doses of low level (i.e., TID < 130 krad).

References

[1] J. D. Cressler, "Using SiGe HBT technology for extreme environment electronics," Proceedings of the Bipolar/BiCMOS Circuits and Technology Meeting, 2005., Santa Barbara, CA, USA, 2005, pp. 248–251, doi: 10.1109/BIPOL.2005.1555243.

[2] F. B. McLean and T. R. Oldham, "Charge Funneling in N- and P-Type Si Substrates," in *IEEE Transactions on Nuclear Science*, vol. 29, no. 6, pp. 2017–2023, Dec. 1982, doi: 10.1109/TNS.1982.4336489.

[3] J. R. Srour, C. J. Marshall and P. W. Marshall, "Review of displacement damage effects in silicon devices," in IEEE Transactions on Nuclear Science, vol. 50, no. 3, pp. 653–670, June 2003, doi: 10.1109/TNS.2003.813197.

[4] Timothy R Oldham. Ionizing radiation effects in MOS oxides, volume 3. World Scientific, 1999.

[5] MIL-STD-883 Test Method Standard, "Microcircuits", Department of Defense, Defense Supply Center, Feb. 22, 2017.

[6] J. D. Cressler, "Radiation Effects in SiGe Technology," in IEEE Transactions on Nuclear Science, vol. 60, no. 3, pp. 1992–2014, June 2013, doi: 10.1109/TNS.2013.2248167.

[7] Z. E. Fleetwood et al., "Advanced SiGe BiCMOS Technology for Multi-Mrad Electronic Systems," in IEEE Transactions on Device and Materials Reliability, vol. 14, no. 3, pp. 844–848, Sept. 2014, doi: 10.1109/TDMR.2014.2331980.

[8] R. Baumann, K. Kruckmeyer, "Radiation Handbook for Electronics", Texas Instruments, http://www.ti.com/radbook.

[9] P. Chevalier et al., "A 55 nm triple gate oxide 9 metal layers SiGe BiCMOS technology featuring 320 GHz fT / 370 GHz fMAX HBT and high-Q millimeter-wave passives," 2014 IEEE International Electron Devices Meeting, San Francisco, CA, USA, 2014, pp. 3.9.1–3.9.3, doi: 10.1109/IEDM.2014.7046978

[10] V. N. Hegde *et al.*, "A Comparison of Electron, Proton and Gamma Irradiation Effects on the I-V Characteristics of 200 GHz SiGe HBTs," in *IEEE Transactions on Device and Materials Reliability*, vol. 18, no. 4, pp. 592–598, Dec. 2018, doi: 10.1109/TDMR.2018.2875064.

[11] J El Beyrouthy, Bruno Sagnes, F Pascal, A B Ayenew, J Boch, et al.. Effects of X-ray and Gamma irradiation on advanced Si/SiGe:C HBTs DC characteristics. *RADiation and its Effects on Components and Systems Conference*, Centre National d'Etudes Spatiales, Sep 2023, Toulouse, France.

[12] A. A. Belie *et al.*, "Low Frequency Noise Study of X-ray Irradiated Si/SiGe:C BiCMOSTechnologyBipolarTransistors," *2023 International Conference on Noise and Fluctuations (ICNF)*, Grenoble, France, 2023, pp. 1–4, doi: 10.1109/ICNF57520.2023.10472764.

[13] J. Zhang *et al.*, "Impact of Bias Conditions on Total Ionizing Dose Effects of 60Coγ in SiGe HBT, in *IEEE Transactions on Nuclear Science*, vol. 63, no. 2, pp. 1251–1258, April 2016, doi: 10.1109/TNS.2016.2522158.

[14] J. M. Roldan, Guofu Niu, W. E. Ansley, J. D. Cressler, S. D. Clark and D. C. Ahlgren, "An investigation of the spatial location of proton-induced traps in SiGe HBTs," in IEEE Transactions on Nuclear Science, vol. 45, no. 6, pp. 2424–2429, Dec. 1998, doi: 10.1109/23.736481.

[15] M.J. Deen and F. Pascal, "Electrical characterization of semiconductor materials and devices-review" *J. Mater. Sci: Mater Electron*, 17, pp 549–575, 2006

[16] M.J. Deen and F. Pascal, "Review of low-frequency noise behaviour of polysilicon emitter bipolar junction transistors", IEE Proc. Circuits Devices Syst, 151,2, p. 125–137, 2004.

[17] F. Pascal et al "Comparison of low frequency noise in III-V and Si/SiGe HBTs", IEE Proc. Circuits Devices Syst, 151,2, p. 138–146, 2004

[18] TARANTO H2020 ECSEL project, co-funding by the French Ministry of Economy, Finance and Industry.

[19] M. Seif, et al "Dispersion study of DC and Low Frequency Noise in SiGe:C Heterojunction Bipolar Transistors used for mm-Wave to Terahertz applications "Microelectronics Reliability (2014) DOI: 10.1016/j.microrel.2014.07.027

[20] J. Elbeyrouthy *et al.*, "Low frequency noise characterization and modeling of SiGe HBT featuring LASER annealing in a 55-nm CMOS node," *2019 International Conference on Noise and Fluctuations (ICNF)*, Neuchatel, Switzerland, Jun. 2019, doi: https://doi.org/10.5075/epfl-iclab-icnf-269248.

[21] C. Mukherjee et al., "Low-Frequency Noise in Advanced SiGe:C HBTs—Part II: Correlation and Modeling," in IEEE Transactions on Electron Devices, vol. 63, no. 9, pp. 3657–3662, Sept. 2016, doi: 10.1109/TED.2016.2588318.C.

[22] Mukherjee *et al.*, "Random telegraph noise in SiGe HBTs: Reliability analysis close to SOA limit," *Microelectronics Reliability*, Jun. 2017, doi: https://doi.org/10.1016/j.microrel.2017.05.001.

B. Sagnes received the Ph.D. degree in components, signals, systems from the University Montpellier 2 in 1992. He was the first to elaborate the CuInSe2 material by OMCVD for photovoltaic applications. He stayed one year at the FORTH in Heraklion, Greece. He studied the MBE elaboration of III-V and SiC devices. From 1993 to 1998, he was an assistant professor in semiconductor physics at the Marseille's engineer school ENSPM. He was interested in MOS technologies reliability. He returned to the University Montpellier 2 in 1998 and studied radiations effects by TCAD simulations, especially single event upset in CMOS SRAM cells. Since September 2010, he has been an assistant professor HDR with the Institut d'Electronique et des Systèmes of the Université de Montpellier. His research interests are is low-frequency noise characterizations of HBTs developed in advanced BiCMOS technologies. His research currently focuses on the effects of X-ray and gamma radiation on these transistors. He has authored or co-authored 98 international publications and conferences.

A. Ayenew received the European master's degree from Ku Leuven (Belgium), the University of Jyvaskyla (Finland), and the University of Montpellier (France) in radiation and its effects on microelectronics devices in 2023. He is currently a Ph.D. student at the Institute of Electronics and Systems (IES), University of Montpellier. His research focuses on radiation effects on the DC characteristics and low-frequency noise of SiGe:C heterojunction bipolar transistors and circuits.

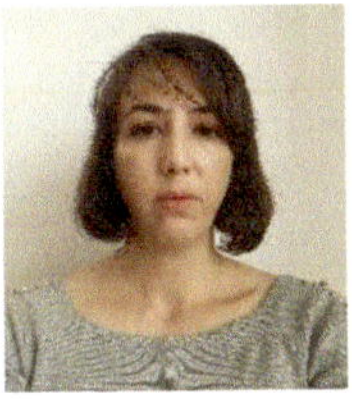

M. Bouhouche received the Ph.D. degree in electronics from the University of Constantine, Algeria, in 2012. She joined the University of Oum El Bouaghi as an assistance professor in 2015. Her area of research interest includes modeling and characterization of HBTs developed in advanced BiCMOS technology. Her current research focuses on effects of radiation on SiGe HBTs, and RF-integrated circuit designed in BiCMOS technology for wireless communications.

J. El Beyrouthy received the master's degree from the Lebanon University in microsystems and embedded electronics in 2017. He received the Ph.D. degree in electronics from the University Montpellier in 2020. His thesis work focused on the characterization and modeling of low-frequency noise in pre- and post-irradiated bipolar transistors developed in BiCMOS technology.

F. Pascal received the M.S. and Ph.D. degrees in electrical engineering from the University of Montpellier, France, in 1986 and 1989, respectively. He was promoted to the rank of Professor in 2004. He is involved in a research team working in electronic devices (MOS, MODFET, MESFET, Si-Ge HBTs, III-V HBTs, etc.) by means of their proper background IV/CV characteristics and noise spectral analysis. He is also working on the technological qualification of contacts, semiconductor materials, and carbon-nanotubes-based devices by the use of noise spectroscopy. He is currently working on low-frequency noise measurement and modeling in advanced SiGe:C heterojunction bipolar transistors in the Institut d'Electronique et des Systèmes, University of Montpellier. His more recent research concerns the effects of X-ray and gamma irradiations on SiGe:C HBTs and associated circuits. He is the author and co-author of more than 120 peer-review publications in international journals and conferences. In 2013 and 2019, he was co-chairman of the 22nd and 25nd editions of the *International Conference on Noise and Fluctuations* and co-editor of the proceedings.

J. Boch received the M.S. degree in 2000 from the Montpellier University and the Ph.D. degree in 2003 in microelectronics from the University of Reims Champagne-Ardenne, France. Prior to accepting a research and teaching position at the University of Nice Sophia-Antipolis, France, in 2003, he spent about a year at Vanderbilt University (USA). In 2004, he joined the RADIAtion and Component (RADIAC) team as a senior lecturer at the University of Montpellier and he is now a full time professor. His research activity has been directed to the study of dose effect in electronics, and reliability of semiconductor devices and systems. He led the RADIAC team from 2016 to 2024. He is now head of the Electrical Engineering Department at the Montpellier University Institute of Technology. He has authored or co-authored over 150 papers in referenced journals or conferences. He has been technical chair at RADECS 2010 in Langenfeld, Austria, and he served as the session chairman at various RADECS conferences.

P. Chevalier (Member, IEEE) received the Ph.D. degree in electronics from the University of Lille, France, in 1998, for his work on InP-based HEMT. He joined Alcatel Microelectronics, Belgium, in 1999, where he contributed to the start of RF BiCMOS. Since joining STMicroelectronics, Crolles, France, in 2002, he has been working on the development of SiGe BiCMOS and RF-SOI CMOS technologies and related devices, with a long-lasting research interest in SiGe HBT. He led the RF-SOI CMOS, BiCMOS & Photonics Technologies R&D team. He is currently Analog/RF Technology Architect and is a Fellow of Technical Staff. Dr. Chevalier has authored or co-authored over 200 technical journal papers and conference publications. He has served on the Technical Program Committees of the IEEE BCTM, the ECS SiGe Symposium, the IEEE BCICTS, and the IEEE IEDM conferences. He has been a member of the RF & AMS Technologies section of the ITRS of which he led the Silicon Bipolar & BiCMOS subgroup.

D. Gloria received in 1995 the engineering degree in electronics from the Ecole Nationale Supérieure d'Electronique et de Radioélectricité and the M.S.E.E. in optics, opto-electronics and micro-waves design systems from the Institut National de Grenoble (INPG).

He spent two years, from 1995 to 1997, in ALCATEL Bell Network System Labs, in Charleroi, Belgium, as an RF designer engineer and was involved in the development of the Cablephone RF front end and its integration in hybrid-fiber-coax telecommunication networks. Since 1997, he has been working within STMicroelectronics, in the Technology Design Platform Department, Crolles. His research interests are in the development and optimization of RF and photonics platforms for applications in BiCMOS and CMOS advanced technologies

D-band Modulated Signal Generation using Photonics Techniques

Y. M. Wodaje, B. Baridi, O. Occello, F. Danneville, P. Szriftgiser, and G. Ducournau

Abstract—**D-band communications are envisioned for the next frequency step forward to support the future back-haul networks. Beyond technology developments that aim to tackle circuits and systems, testing approaches also need to support the frequency upgrade of systems. This paper presents the photonics approach to generate modulated signals in the range 110–170 GHz. Typical performances that can be reached using standard lasers are discussed, as well as the measurement of error vector magnitude induced penalty in relation to the power-compression curve of a D-band amplifier under test.**

Index Terms—**Millimeter-waves, THz measurements, THz communications**

The work received support from Contrat de Plan Etat Region (CPER) WAVETECH @HdF project and Haut de-France Regional council, IEMN UHD flagship project, the IEMN CHOP platform, France 2030-PEPR FUNTERA and SYSTERA projects, funded by ANR (Agence Nationale de la Recherche) under grant agreement numbers ANR-22-PEEL-0006 and 22-PEFT-0006, French RENATECH network, the Equipex + Nanofutur (grant agreement number IA-21-ESRE-0012). D-band testbed development is also supported by the SHIFT project. SHIFT is supported by the Chips Joint Undertaking and its members, including the top-up funding by National Authorities, under grant agreement number 101096256. The work is also supported by the IPCEI Microelectronics and Communication Technology (ME/CT), IPCEI-ME-CT, Project IEMN-SP05, and BPI at French side.

Yewulsew Manale Wodaje, Bachar Baridi, François Danneville, and Guillaume Ducournau (guillaume.ducournau@univ-lille.fr) are with Université de Lille, CNRS, Centrale Lille, Université Polytechnique Hauts-de-France, UMR 8520-IEMN, Institut d'Electronique de Microélectronique et de Nanotechnologie, F-59000-Lille, France.

P. Szriftgiser is with Université de Lille, CNRS, UMR 8523 – PhLAM, Laboratoire de Physique des Lasers, Atomes et Molécules, F-59000-Lille, France.

I. INTRODUCTION

For several years, the development of active circuits for amplification in the 140–170 GHz range has progressed significantly, requiring corresponding advancements in characterization tools. In addition to conventional linear characterization, linearity measurements using dual-tone signals and wide-band signal are imperative to accurately evaluate the system-level performance of amplifiers. This paper briefly describes the development of a modulated signal analysis (MSA) technique for D-band measurements. This paper is supported by a presentation titled "Key enabling technologies for future wireless, wired, optical and satcom applications (KETCOM)" in an international workshop at the European Microwave week (EuMW, www.eumweek.com), Paris, France, on September 22, 2024. This workshop captures the latest research roadmaps and achievements on these topics from the European ecosystem, comprising industry, research, and academia.

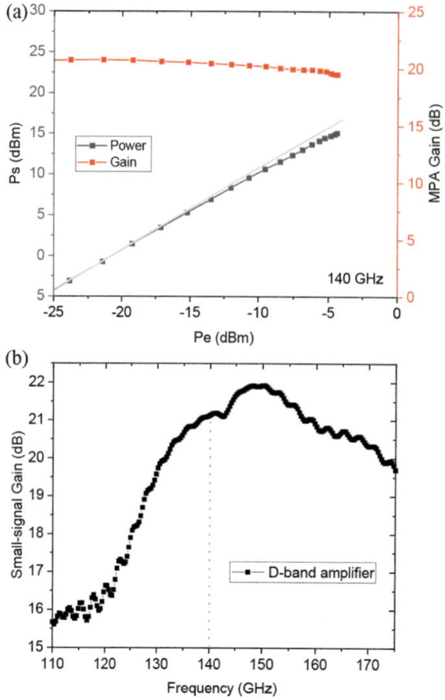

Fig. 1. (a) Small-signal gain of the DUT, in the whole WR6.5 (110–170 GHz) band. (b) DUT power curve at 140 GHz.

II. MODULATED SIGNAL ANALYSIS OF D-BAND AMPLIFIER

This section describes our investigation of MSA for the D-band measurements. In this investigation, we utilize the D-band transmitter (Tx) source established in the Chips JU project SHIFT. With this transmitter, we test the MSA using a non-linear circuit. This circuit is a medium power amplifier (MPA) working in the D-band. Figure 1a illustrates the typical gain of the device using the III-V technology. Figure 1b presents a typical compression curve of the amplifier at 140 GHz.

One of the main key performance indicators of an amplifier is its linear power level available throughout its operational lifetime. The P_{1dB} metric is typically used to evaluate the linearity of the circuit. Alternatively, the third-order intermodulation metric is used. The saturated power (P_{SAT}) is sometimes reported, but this provides only initial insights into amplifier linearity. In our study, the P_{1dB} was characterized and found to be approximately 12.5 dBm. In practical telecommunication scenarios such as back-hauling applications, amplifiers integrated into analog front-ends must handle advanced modulation formats like quadrature amplitude modulation (QAM). This necessitates a power back-off from P_{1dB} to achieve fully linear operation. This required back-off can vary depending on the circuit architecture and the specific modulation format employed.

III. MODULATED SIGNAL ANALYSIS SETUP AND MEASUREMENTS

For telecommunications applications, amplifier properties must often be evaluated over the entire channel defined by the standard. Consequently, new testbeds are required based on wideband signals instead of continuous-wave single tones. To assess the performances of the MPA for communication applications, we propose using a complex modulated signal injected into the device under test (DUT). At amplifier's output, any non-linear effect will degrade the quality of the constellation, which can be qualified by the error vector magnitude (EVM, %). The variation of the EVM with output power is an effective method for assessing the amplifier linearity, as EVM should remain constant if the amplifier is within in its linear regime.

To this end, we employ a reference transmitter (Tx in Figure 2) that generates a complex wideband modulation with a 140 GHz carrier frequency and a calibrated input power (P_{in}). The system uses optical techniques in coherent optical

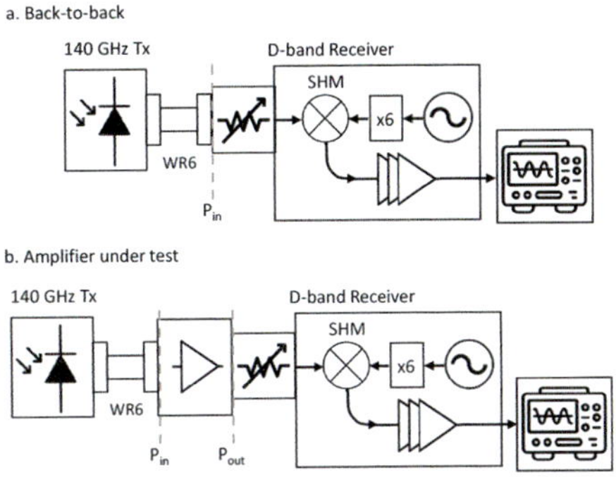

Fig. 2. (a) Back-to-back measurement in MSA analysis. (b) Evaluation of the EVM obtained using the amplifier under-test. SHM: Sub-harmonic mixer is used to down-convert the signal from 140 GHz to an intermediate frequency, further detected on a w wideband real-time oscilloscope. The attenuator is used to optimize the power injected into the receiver.

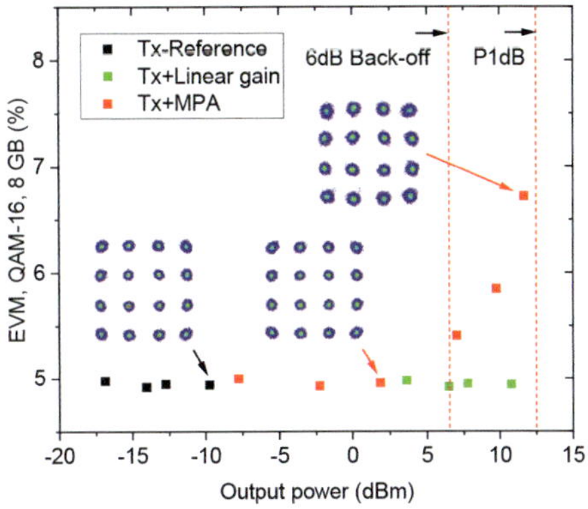

Fig. 3. EVM performances at 140 GHz. EVM of the reference Tx combined with the receiver (black). Expected EVM extrapolation of the amplifier linear gain at 140 GHz (green). Actual performance of the system with the amplifier under test (red). Dashed lines show the P_{1dB} and 6 dB back-off from P_{1dB}.

fiber communication systems. Initially, a set of two laser lines is used: one is I/Q modulated using a Mach-Zehnder modulator, and the other remains a continuous wave. These two lasers are coupled and amplified before being injected into the fast photodiode (Tx). A photomixing process occurs in at the fast photodiode, where the two lasers, detuned by 140 GHz, produce a D-band modulated signal at the photodiode. By utilizing an arbitrary waveform generator to create the I/Q base-band signals, both the modulation format and the baud rate can be adjusted. In this example, an 8 GBaud rate with 16-state QAM-16 was considered, resulting in a data rate of 32 Gbit/s. The main advantage of photonics over fully electronics techniques is the ease of modulation at the optical level and the frequency tenability, which can span the entire D-band.

This modulated signal is then used a test signal and the EVM evolution versus output power is shown in Figure 3. The black curve presents the EVM of the reference transmitter, while the green curve shows the expected EVM with a linear amplifier using the small signal gain illustrated in Figure 1. The red curve depicts the actual performance of the system with the amplifier under test. As expected, the red curve is shifted to the right, indicating that the MPA can provide more power than the reference Tx source. However, beyond 5 dBm at MPA output, a noticeable EVM degradation occurs before reaching the DUT's P_{1dB}. In this configuration, maintaining less than 0.5% EVM degradation requires a 6 dB back-off from P_{1dB}, while a 10 dB back-off from P_{1dB} results in almost no degradation.

IV. CONCLUSION AND OUTLOOK

Next, we plan to optimize the reference EVM levels obtained from the transmitter and study the transmitter's linearity to determine available linear range if P_{in}. Additionally, the capability of the developed testbed over the upper part of the D-band (from 140 to 170 GHz) will be evaluated. The MSA technique will then be applied to assess the performance of D-band amplifiers using BiCMOS technology, designed and fabricated in the Chips JU project SHIFT.

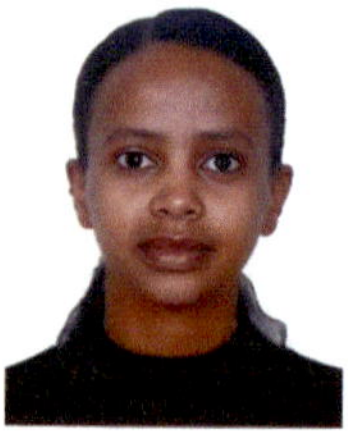

Yewulsew Manale Wodaje was born in Bahir Dar, 1994, Ethiopia. She graduated from Bahir Dar Institute of Technology, Bahir Dar University (BiT), Bahir Dar, Ethiopia, in 2018 (B.Sc. degree in electronics and communication engineering), in 2023 (M.Sc. degree in microelectronics engineering) and started working toward her Ph.D. degree at the Institute of Electronics, Microelectronics and Nanotechnology (IEMN), UMR-CNRS 8520, University of Lille, Villeneuve d'Ascq, France, since January 1, 2024 on the thesis entitled " Contribution to circuit design and advanced characterization for D-band communications applications".

Bachar Baridi received the master's degree in microelectronics from the Univeristé de Strasbourg, France, in 2023. His research interests cover the power amplifiers and linearization techniques. He is now working toward a Ph.D. degree at the Institute of Electronics, Microelectronics and Nanotechnology (IEMN), UMR-CNRS 8520, University of Lille, Villeneuve d'Ascq, France.

Olivier Occello received the M.Sc. degree in wireless integrated circuits and systems from Grenoble Alpes University, France, and the Ph.D. degree from Université Grenoble Alpes as a part of TIMA, Grenoble, France, in 2024. He is currently with IEMN, University of Lille/CNRS, France. His research interests include the design of mmWave integrated circuits and their built-in tests.

François Danneville received the Ph.D. degree in electronics and the Habilitation à Diriger les Recherches from the Lille1 University of Sciences and Technology, Villeneuve-d'Ascq, France, in 1991 and 1999, respectively. Until 2000, he was an assistant professor with Lille1 University. He has studied the noise of III–V devices operating in linear and non-linear operations (in LNA, mixer, and oscillator). In 1998, on the aforementioned research topic, he made one year sabbatical as a visitor scientist in HP (nowadays Keysight Technologies), Santa Rosa, CA, USA. Since 2001, he has been a full professor with Lille University, where his research has focused on the small (large) signal and noise properties in high-frequency range (centimeter and millimeter waves) of advanced silicon technologies (CMOS and SiGe HBTs), including alternative architectures and circuit design. Since 2014, his research work has been focusing on the development of an ultralow-power silicon CMOS artificial neuron (published in *Frontiers in Neuroscience*, March 2017) and is currently focusing on the integration (scaling) of this new neuromorphic technology within neural networks. His research areas of interest (to cite a few) are biocomputing, smart visual and hearing sensors, electronic systems for neural recording and stimulation, and fluctuations and noise in artificial and biologic neural systems.

Pascal Szriftgiser received the Ph.D. degree in quantum physics from the Laboratoire Kastler Brossel, University Paris VI (Sorbonne University today), Paris, France, in 1996. The same year, as a researcher, he joined CNRS (Centre National de la Recherche Scientifique) at Laboratory PhLAM (Physique des Lasers Atomes et Molécules). Since 2012, he has been CNRS Research Director. From 2014 to 2021, he is the Leader of the Cold Atoms Group and Deputy Director of the Laboratory PhLAM from 2019 to 2021. Since 2021, he has been a PI of Contrat de Projet Etat Région (CPER) WaveTech@HdF, a consortium that brings together six research laboratories in Hauts-de-France country. He has authored or coauthored 137 publications in peer-reviewed international journals or conference proceedings. His expertise spans from cold atoms physics, quantum chaos, topology, and non-linear optics in optical fibers to high-speed coherent optical communications for THz applications.

Guillaume Ducournau was born in Clamart, France, in 1979. He graduated from ESIGELEC, Rouen, France in 2002 (Engineering degree), and received the Ph.D. degree from the University of Rouen, France, in 2005, on optical fiber communications. He has been with the Institute of Electronics, Microelectronics and Nanotechnology (IEMN), UMR-CNRS 8520, University of Lille, Villeneuve d'Ascq, France, since 2007. He is the Leader of the THz wireless communications activity with IEMN using optoelectronic THz photomixers, electronic receivers, THz instrumentation, and millimeter-wave (mmWave) characterization. He worked on several European projects: STREP ROOTHz 2010–2013, THoR H2020, GRAPH-X, TIMES (6G SNS), as well as the Marie-Curie TERAOPTICS network. At national level, he was the coordinator of the COM'TONIQ Project from 2014 to 2017 funded by ANR (INFRA 2013) dedicated to THz communications in the 300 GHz band, the ANR/DFG TERASONIC project for the use of THz photonics technologies and electrical solid-state technologies for THz communications, and SPATIOTERA for spatially distributed photomixers. He received the 2020 ISAP BEST PAPER award. He is involved in national France 2030 programs gathering several French laboratories under the "PEPR" programs funded by the ANR (Agence Nationale de la Recherche). In this framework, the FUNTERA project (6 partners) is investigating THz converters, while the SYSTERA project (12 partners) is dedicated to beyond 90 GHz systems for future networks. He also participates to the ST-IEMN common laboratory, and more specifically involved in the mmWave technologies characterization part, as in the JU SHIFT European program. He has authored or co-authored more than 180 publications in peer-reviewed international journals or peer-reviewed conferences proceedings and holds one patent. Professor Ducournau is serving as General Chair of the European Microwave Week in Paris, 2024, "Waves Connecting Europe."

Decarbonizing the Electronics Industry to Achieve Net Zero (2024)

Gunther Walden

Abstract—**The electronics industry, a vital pillar of modern society, faces substantial challenges in its path to achieving net-zero emissions by 2050. Central to this effort is the need to decarbonize the entire supply chain. This article explores the complexities associated with collecting primary emissions data from both manufacturing processes and supply chain partners. It emphasizes the importance of primary data over secondary data, which only provides industry averages that are insufficient to achieve a net-zero target. The discussion looks at technological advancements, collaborative strategies, and policy measures required to transform the electronics industry into a net-zero industry.**

Index Terms—**Carbon emissions, carbon footprint, carbon neutral, environmental, supply chain, sustainable development.**

I. INTRODUCTION

THE electronics industry has become an integral part of modern life. It is the basis for the operation of smartphones, computers, medical devices, and numerous other technologies. And its dissemination is expected to further increase, according to the European Chips Report, the future chip demand across industrial ecosystems is expected to double between 2022 and 2030 [1]. However, the manufacturing processes, the extensive supply chain, and the electricity demand during the use phase generate high levels of carbon emissions. As part of the global community's efforts to combat climate change, the electronics industry faces the daunting task of achieving net-zero emissions by 2050. This goal requires a comprehensive approach that includes reducing emissions in manufacturing, improving transparency in the supply chain, and innovating energy-efficient technologies. The focus of this article is how transparency about supply chain emissions can be achieved and how supply chain emissions reduction can be managed.

A. Understanding Net-Zero Targets and Their Meaning

Net-zero targets refer to the balance between the amount of greenhouse gases (GHG) emitted into the atmosphere and the amount that is removed or offset, resulting in a net-zero carbon footprint. These targets are critical to mitigating climate change and preventing global temperatures from rising to dangerous

This work received funding from Chips Joint Undertaking (Chips JU) under grant agreement number 101096256 (project SHIFT). The Chips JU receives support from the European Union's Horizon Europe research and innovation program and the National Authorities.

Gunther Walden (gunther@circulartree.com) is with CircularTree, Germany, a company that envisions a future where businesses seamlessly integrate sustainability into their operations.

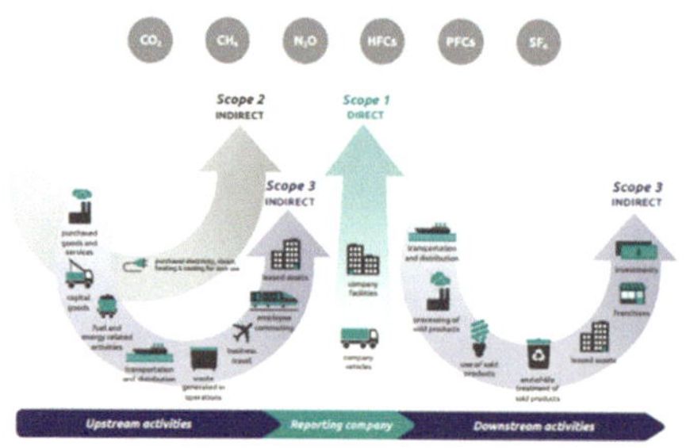

Fig. 1. Overview of scopes and emissions across a value chain [3].

levels. Achieving net zero requires reducing emissions as much as possible and offsetting the remaining emissions through compensation strategies such as reforestation or carbon capture and storage. Net-zero targets are important because they provide governments, businesses, and organizations with a clear and measurable goal to work toward to ensure that efforts to combat climate change are coordinated and effective. By committing to net-zero emissions, companies demonstrate their commitment to sustainability, reducing the impact of global warming and protecting the environment for future generations [2].

"The Greenhouse Gas Protocol" supplies the most widely used GHG accounting standards and guidance. It defines which GHGs are to be considered and structures GHG emissions in direct emissions (Scope 1), indirect energy related emissions (Scope 2), and indirect supply chain emissions (Scope 3 upstream and Scope 3 downstream) [3]. See Figure 1 for the overview of scopes and emissions across a value chain.

Theterms"carbonneutral"and"net-zero" are often used interchangeably, but they have distinct meanings and implications.

Carbon neutrality typically covers only a defined part of business operations, e.g., Scope 1 and Scope 2, and means that any carbon dioxide released into the

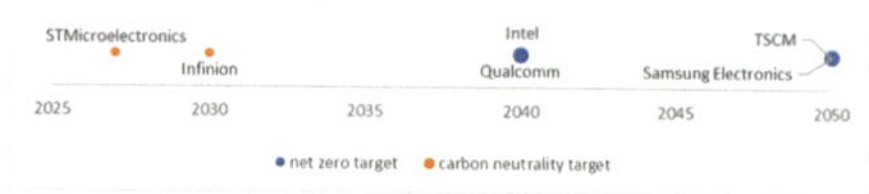

Fig. 2. Net-zero/carbon neutrality target dates of selected chip manufacturers [5][6][7][8][9][10].

atmosphere from a company's activities is balanced by an equivalent amount being removed or offset. This can be achieved by purchasing carbon credits that support renewable energy projects, reforestation, or other carbon offset initiatives.

Net-zero emissions means that a company includes all its GHG emissions across its whole supply chain (Scope 1, Scope 2, and Scope 3) to as close to zero as possible, and any remaining emissions are balanced out by removing an equivalent amount of GHGs from the atmosphere. [4]

Figure 2 illustrates the climate targets of some leading semiconductor manufacturers as of May 2024. Several companies are aiming for a net-zero target for 2050. Others have committed to carbon neutral goals.

B. Importance of Primary Data

Secondary data related to GHG emissions refers to information previously collected and published by other sources. This data is usually aggregated and available through various platforms, including industry reports, government publications, and foremost life cycle assessment databases such as Ecoinvent and GaBi. Secondary data provides industry averages, general trends, and benchmark values in terms of GHG emissions from different sectors, processes, and activities. Although widely used, it does not capture the specific emissions of individual suppliers, processes, or products. Therefore, while secondary data can be valuable to understand general patterns and establish a baseline, it is insufficient to achieve precise and targeted emission reductions required for net-zero targets.

Primary data is company-specific data collected from individual suppliers. Accurate primary data is essential for identifying emission hotspots and implementing targeted reduction strategies [11].

C. Corporate Carbon Footprint vs. Product Carbon Footprint

When accounting for GHG emissions, a distinction is generally made between two basic approaches: The corporate carbon footprint describes the GHG emissions of a company as a whole. The product carbon footprint (PCF), on the other hand, comprises product-specific emissions, i.e., emissions that can be directly attributed to a product [11]. To calculate carbon emissions along the supply chain, it is required to exchange PCF at every step of the supply chain.

II. EMISSIONS FROM OWN MANUFACTURING

The first major challenge in decarbonizing the electronics industry is for companies to accurately measure and reduce emissions from their own manufacturing processes. These emissions come from various sources, including the direct consumption of fossil fuels and electricity consumption.

A. Energy Consumption in Manufacturing

Production facilities in the electronics industry are generally highly automated and rely heavily on electricity. Especially cleanrooms consume large amounts of energy compared with non-classified rooms; scientific literature and experience in the field show that cleanrooms use up to 25.3 times more energy (1.25 kW/sqm vs. 0.06 kW/sqm). The energy requirement of HVAC systems usually amounts to 50%–75% of electricity consumption in a clean production space due to the high airflow rates needed for particular ISO classes [12]. But also processes such as soldering, etching, and assembly are energy-intensive. The production of printed circuit boards (PCBs), for example, involves several steps that consume energy, such as laminating, drilling, and coating.

B. Strategies for Reducing Emissions

Technological Advancements in Manufacturing

To mitigate these emissions, the industry is exploring various technological advancements. These include the development of more energy-efficient fabrication techniques, e.g., STMicroelectronics has reduced its energy consumption per unit of production by 56% since 1994 [13].

Renewable Energy Sources

The use of renewable energy sources is a widely used method to reduce carbon emissions. Apple, for example, has committed to powering all of its facilities with 100% renewable energy and has already achieved this milestone for its corporate offices, data centers, and retail stores worldwide [14]. However, it is important to understand that due to the fact that Apple's manufacturing is outsourced to third-party suppliers, the manufacturing facilities are not included in that renewable energy claim.

Product Design

In many products, the rigid PCB also primarily fulfills a mechanical function, which simplifies the product design but worsens the GHG balance. Sometimes up to half of the PCB area is used merely to bridge a gap between two pole connections, for example. Alternative approaches to secure fixing and contacting could significantly improve the GHG balance.

Advanced Cooling Technologies

Innovation in advanced cooling technologies is another key area. Traditional cooling methods for semiconductor manufacturing facilities are energy-intensive. Newer methods, such as immersion cooling and more efficient HVAC systems, can significantly reduce energy consumption.

Materials Science Innovations

Advancements in materials science are also contributing to lower emissions. Researchers are developing new materials that require less energy to produce and have better electrical properties, which can reduce the overall energy consumption of electronic devices. For instance, gallium nitride semiconductors are more efficient than traditional silicon semiconductors and are increasingly used in power electronics.

III. SUPPLY CHAIN EMISSIONS

In order to accurately calculate the environmental impact of electronic products, it is essential to determine the PCF. This calculation requires a comprehensive analysis of the upstream supply chain for all suppliers and sub-suppliers. The different stages of the upstream supply chain are referred to as Tier 1, Tier 2, Tier 3, etc., starting from the original equipment manufacturer. Figure 3 shows an example of an upstream supply chain.

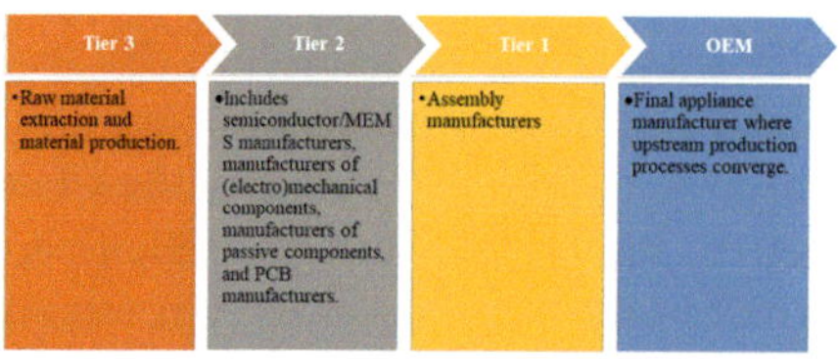

Fig. 3. Exemplary upstream supply chain.

Understanding these stages and the associated emissions is crucial for calculating the PCF and achieving net-zero targets.

A. Complexity and Lack of Standardization

Collecting primary emissions data from the supply chain is a complex challenge. The electronics supply chain is global and involves multiple tiers of suppliers spread across different regions with different environmental regulations and levels of transparency. The lack of standardized calculation and reporting practices increases the difficulty of obtaining accurate emission data. If PCF calculations are not comparable between different manufacturers and if there is not a standardized reporting template, the efforts for companies along the supply chain are tremendous; that is why as of today, only very few companies actually provide PCF data.

B. Initiatives to Improve Transparency

Partnership for Carbon Transparency

The Partnership for Carbon Transparency (PACT), initiated by the World Business Council for Sustainable Development aims to enhance transparency and decarbonization in the supply chain by enabling the secure and standardized exchange of product-level emissions data across organizations. The Pathfinder Framework [15] developed by PACT provides guidelines for calculating and exchanging primary-data-based PCFs, improving data reliability and consistency across value chains. This initiative also involves collaboration with technology providers like Siemens, SAP, and CircularTree to create interoperable solutions for data exchange, thus fostering transparency and accountability in corporate climate action.

Industry Initiatives

Addressing the emission challenges in the electronics industry requires a collaborative approach. Companies must work closely with their supply chain partners to ensure transparency and consistency in emission reporting. Industry-wide initiatives, such as those led by the Japan Electronics and Information Technology Industries Association (JEITA), are crucial for fostering cooperation and setting standardized practices across the sector. JEITA focuses on promoting sustainable development and environmental management in the electronics industry, advocating for standardized reporting practices and collaborative efforts to reduce emissions throughout the supply chain.

C. Workflow for PCF Transparency and Reduction

CARE Decarbonization Cycle

The key to practical transparency and decarbonization is efficiency. Only if companies, especially small- and medium-sized companies, get the opportunity to achieve transparency and decarbonization with a reasonable effort, the targets can be achieved. To empower companies to do this, CircularTree has developed the CARE decarbonization cycle. This approach enhances supply chain transparency and facilitates a gradual path to decarbonization. The process begins with uploading a bill of material and a hotspot analysis using secondary data to identify key suppliers and components with significant emissions. These identified suppliers are then prioritized for requesting primary data. By obtaining and utilizing this primary data, reduction targets can be established collaboratively. This method continuously refines the PCF by

Fig. 4. CARE decarbonization cycle.

progressively replacing secondary data with primary data and implementing agreed-upon reduction targets. As the primary data share increases, the accuracy and value of the PCF improves. The steps of the *CARE* decarbonization cycle as illustrated in Figure 4 include:

1. *Calculate* the PCF, initially based on secondary data
2. *Analyze* emission data to identify hotspot suppliers
3. *Report* the PCF to customers *Request* emission data from hotspot suppliers
4. *Engage* with suppliers and agree on reduction targets

IV. POLICY MEASURES

A. Regulatory Frameworks

Government policies play a crucial role in driving the decarbonization of the electronics industry. Regulations that mandate emission reporting and set stringent emission reduction targets can compel companies to adopt greener practices. For instance, the European Union's Green Deal aims to make Europe climate neutral by 2050 and includes measures to reduce emissions from the electronics sector.

Incentives and Penalties

Incentives for using renewable energy and penalties for high carbon footprints can further accelerate the transition to net zero. Governments can provide tax breaks, grants, and subsidies for companies that invest in renewable energy projects and energy-efficient technologies. Conversely, imposing carbon taxes on high-emission activities can encourage companies to reduce their carbon footprint.

B. International Agreements

International agreements, such as the Paris Agreement, also influence the electronics industry's approach to decarbonization. These agreements set global targets for emission reductions and encourage countries to implement policies that support sustainable practices. Companies operating internationally must comply with these regulations and align their strategies with global sustainability goals.

V. CONCLUSION

Achieving net-zero emissions in the electronics industry by 2050 is crucial due to its significant environmental impact. A multitude of strategies by all companies in the industry needs to be put in place to achieve this target. It requires technological innovation, both in product design and manufacturing technology, as well as improved supply chain transparency with the widespread use of primary data.

Primary data is essential for pinpointing emission hotspots and implementing effective reduction strategies. Advanced technologies and energy-efficient processes are key to reducing manufacturing emissions.

The global and complex nature of the electronics supply chain necessitates initiatives like the PACT, which enhances data transparency and consistency by setting the appropriate standards. Industry-specific collaborations and standardized reporting practices are vital for achieving reliable emissions data.

Policy measures, including regulatory frameworks, incentives, and international agreements like the Paris Agreement, support these efforts by creating an environment that encourages sustainable practices.

In summary, the electronics industry can achieve net-zero emissions through coordinated efforts that leverage technological advancements, supply chain transparency with primary data and supportive policies, contributing to a sustainable future.

REFERENCES

[1] European Comission, (2022). European Chips Report. [Online]. Available: https://single-market-economy.ec.europa.eu/industry/strategy/digital-transformation/european-chips-report_en#.

[2] United Nations. (2023). Climate Action – Net Zero Coalition. [Online]. Available: https://www.un.org/en/climatechange/net-zero-coalition#.

[3] World Resources Institute and World Business Council for Sustainable Development. (2004, March). The Greenhouse Gas Protocol - A Corporate Accounting and Reporting Standard. [Online]. Available: https://ghgprotocol.org/sites/default/files/standards/ghg-protocol-revised.pdf.

[4] World Economic Forum. (2022, August 23). Carbon neutral and net zero – what do they mean?. [Online]. Available: https://www.weforum.org/agenda/2022/08/carbon-neutral-net-zero-sustainability-climate-change/.

[5] STMicroelectronics. (2020, Dec. 09). STMicroelectronics to be Carbon Neutral by 2027. [Online]. Available: https://newsroom.st.com/media-center/press-item.html/c2974.html.

[6] Infineon Technologies AG. (2020, Feb. 20). Infineon to become a carbon-neutral company. [Online]. Available: https://www.infineon.com/

cms/en/about-infineon/press/press-releases/2020/INFXX202002-030.html?redirId=122056.

[7] Intel Corporation. (2023, November 14). Intel Commits to Net-Zero Greenhouse Gas Emissions in its Global Operations by 2040. [Online]. Available: https://www.intel.com/content/www/us/en/newsroom/news/net-zero-greenhouse-gas-emissions-operations.html

[8] Qualcomm Technologies, Inc.. (2021, Oct 31). Qualcomm Announces Goal to Achieve Net-Zero Emissions by 2040. [Online]. Available: https://www.qualcomm.com/news/releases/2021/11/qualcomm-announces-goal-achieve-net-zero-emissions-2040.

[9] Taiwan Semiconductor Manufacturing Company Limited: (2023, September 15). TSMC Accelerates Renewable Energy Adoption and Moves RE100 Target Forward to 2040. [Online]. Available: https://pr.tsmc.com/english/news/3067.

[10] SAMSUNG. (2022, September 15). Samsung Electronics Announces New Environmental Strategy. [Online]. Available: https://news.samsung.com/global/samsung-electronics-announces-new-environmental-strategy.

[11] World Business Council for Sustainable Development. (2023, January). Partnership for Carbon Transparency - Guidance for the Accounting and Exchange of Product Life Cycle Emissions. [Online]. Available: https://wbcsd.github.io/tr/2023/framework-20232601/framework.pdf.

[12] HPCi Media Limited. (2019, Jan. 31). Five forces driving energy efficiency in next-gen cleanroom design. [Online]. Available: https://cleanroomtechnology.com/five-forces-driving-energy-efficiency-in-next-gen-cleanroom-design-151385#.

[13] STMicroelectronics. (2024). We create Sustainable technology in a sustainable way. [Online]. Available: https://www.st.com/content/st_com/en/about/sustainability/sustainable-technology.html.

[14] Apple Inc.. (2018, April 9). Apple now globally powered by 100 percent renewable energy. [Online]. Available: https://www.apple.com/newsroom/2018/04/apple-now-globally-powered-by-100-percent-renewable-energy/

[15] World Business Council for Sustainable Development. (2023, January). PACT: Pathfinder Framework - Guidance for the Accounting and Exchange of Product Life Cycle Emissions. [Online]. Available: https://wbcsd.github.io/tr/2023/framework-20232601/framework.pdf.

[16] CircularTree. (2024). Decarbonization with CircularTree - Reach your net zero goals and optimize your product carbon footprint. [Online]. Available: https://circulartree.com/

Gunther Walden received the engineering degree from the Karlsruhe Institute of Technology, Germany, in 1989. He joined Siemens AG in 1989, where he held numerous positions in Germany and the US, among others being responsible for automotive and fast moving consumer goods industries. He co-founded and has been leading CircularTree as CEO since 2018. He was also a guest lecturer at the Berlin University of Applied Sciences and has held Advisory Board positions for start-ups.

Analog Multiplexing for Bandwidth and Sampling Rate Multiplication of Digital–Analog Converters in Coherent Optical Transmission Systems

Oliver Hauck*, Fred Buchali†, Michael Collisi*, and Michael Möller*, *Member, IEEE,*

*Chair of Electronics and Circuits, Saarland University, Saarbrücken, Germany
†Nokia Network Infrastructure, ION, Stuttgart, Germany

Abstract—**This paper presents the concept of analog time-division multiplexing of DAC output signals and introduces a frequency-domain explanatory model for both 2:1 and 4:1 analog multiplexer (AMUX) variants. Also, it presents 2:1 and 4:1 AMUX circuit concepts and related design considerations along with measurement and simulation results of a 2:1 AMUX module, realized in SiGe BiCMOS technology exhibiting an effective resolution of over 7 bit and a maximum sampling rate of 186 GS/s.**

Index Terms—**Analog multiplexer, AMUX, ENoB, SiGe-BiCMOS, DAC, high speed, 4:1 AMUX**

I. INTRODUCTION

COHERENT optical transmission systems are pivotal for transporting aggregated data traffic in metro, long-haul, and submarine domains. These systems achieve significantly higher transmission rates compared to other parts of the communication network, primarily due to their utilization of extremely high transmission bandwidths combined with high spectral efficiencies. Currently, systems with approximately 130 GHz bandwidth are being introduced to the market [1], supporting bit rates of up to 1.2 Tb/s for long-haul transport and 800 Gb/s as pluggables for metro transport. Such high transmission bandwidths necessitate analog interfaces with electrical bandwidths exceeding 65 GHz. On the transmitter side, these analog interfaces comprise digital-to-analog converters (DACs), driver amplifiers, and modulators, while on the receiver side, they include photodiodes, transimpedance amplifiers, and analog-to-digital converters.

In this paper, we focus particularly on the DAC within the transmitter. In state-of-the-art coherent transmission systems, these converters are implemented using complementary metal–oxide–semiconductor (CMOS) technology and are monolithically integrated with the digital signal processor. The realization of DACs employs a parallelization approach, where the output signals from parallel DACs are gradually serialized using analog multiplexers (AMUX), which are also implemented in CMOS. The achieved bandwidths and spectral efficiencies are relatively high, supporting the elevated data rates.

Currently, discussions are underway regarding the next generation of products, which aim to support transmission rates of 1.6 Tb/s [2]. These advanced systems will require a substantial increase in symbol rates from 130 to 240–260 GBaud. Consequently, the necessary electrical bandwidths will need to be increased to approximately 120–130 GHz. The realization of DACs with such bandwidths presents a significant challenge. Based on current architectures with parallel DACs and gradual serialization, this would involve both increased parallelization and extending serialization to much higher bandwidths at the output stage.

In addition to the monolithic integration of multiplexers in CMOS, there is intensive research into such components based on indium phosphide (InP) [3]–[9] and silicon-germanium [10]–[15] technologies. These multiplexers could be hybrid integrated with the DACs, thereby alleviating the demands on CMOS converters. The AMUXs must support both high bandwidths well beyond 100 GHz and high spectral efficiencies. Concepts such as time-domain multiplexing and frequency-domain multiplexing have been investigated, alongside parallelization as a critical parameter.

The hybrid integration of DACs and AMUXs requires a parallel data interface with two or four parallel data lines and a corresponding clock interface. In most publications, 2:1 multiplexing concepts are applied [7]–[15], where the required clock frequency for interfacing with CMOS application-specific integrated circuits (ASICs) would be extremely high. Alternatively, clock multipliers can be integrated into the AMUX [5] or greater parallelization can be chosen.

In this paper, we follow the idea of greater parallelization and investigate the extension of the 2:1 AMUX principle to a 4:1 AMUX. In Section II, a frequency-domain explanatory model based on the composition of the AMUX spectrum is introduced. Based on that model, the N-fold Nyquist frequency and the origin of unwanted signals are explained. Additionally, basic AMUX circuit concepts are discussed regarding their performance. For the most promising concept, a new extended 4:1 AMUX concept is proposed. Finally, Section IV shows new simulations and measurement results of a recent 2:1 AMUX realization.

II. FREQUENCY-DOMAIN EXPLANATORY MODEL OF THE AMUX

Figure 1a shows the operating principle of an N:1 AMUX, where a clock toggled switch periodically selects each of its N DAC input signals $x_n(t)$ during a sample time $T_{s,a}$, which is $1/N$ of each DAC's sample time $T_{s,d}$. Thereby the sample rate of the AMUX output signal $y(t)$ is N times higher than the DAC sample rate, i.e., $f_{s,a} = N f_{s,d}$. As shown in the following, by the N-fold sampling rate also the Nyquist frequency at the AMUX-output signal is N times the Nyquist frequency of the DAC input signals. Such DAC-AMUX setups are also called AMUX-DAC [16] or Super-DAC [17].

Figure 1a directly shows how the N-fold sample rate of the AMUX is achieved; however, it is not helpful to understand how the N-fold bandwidth is composed. Also, it neither shows how the signal spectrum of the AMUX is created nor how spectral noise impairs the AMUX's effective number of bits given by ENoB = (SINAD − 1.76)/6.02. To point out

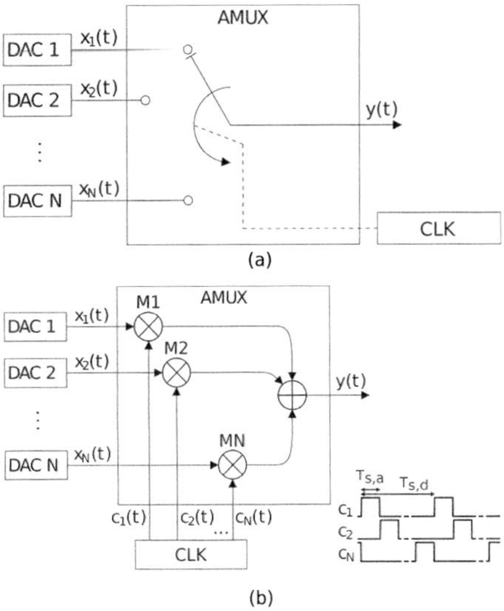

Fig. 1. (a) N:1 AMUX principle as a clocked toggle switch. (b) Illustration of the N:1 AMUX as mixers with added output [16]. In case of the 2:1 AMUX, only two, for the 4:1 AMUX, four mixers are present.

the origin of the main contributors to AMUX noise determined by the signal to noise and distortion ratio (SINAD), it is advisable to introduce a frequency-domain explanatory model of the AMUX, previously presented in [11] for a 2:1 AMUX, and to show how to expand this model to a 4:1 AMUX. As stated in [11], the effective resolution of the AMUX is mainly determined by three major impairments:

1. Since the AMUX-output spectrum is composed of the DAC spectra, the individual DAC's SINAD and therefore their ENoB strongly determines and limits the AMUX's ENoB.
2. Any non-linearity of the AMUX circuitry degrades the SINAD and ENoB by adding harmonics to the AMUX spectrum.
3. As shown by the frequency-domain explanatory model, the ENoB of the AMUX strongly depends on (a) a proper timing of the DAC signals and (b) identical DACs with identical signal paths up to the AMUX multiplexing core. In practical designs it turned out that there are multiple sources of impairments of (a) and (b) that are prone to severely degrade the AMUX's ENoB.

The frequency-domain explanatory model is based on the alternative AMUX block diagram in Figure 1b. There, the selection of the nth DAC signal, as shown in Figure 1a, is represented by the multiplication of the signal by a sampling signal $c_n(t)$ with a non-zero value during selection time. In the case of an ideal selection, $c_n(t)$ is periodic in $T_{s,d}$ and toggles between 1 (selection phase, i.e., sample time $T_{s,a}$) and 0 (not selected).

Ideally, the sampling signals are shifted and aligned in time, such that the individual DACs are evenly and sequentially selected. In case of a 2:1 AMUX, there are only two sampling signals $c_{1,2}(t)$, toggling between 1 and 0 with a 1:1 mark-space ratio and shifted by 180°. Due to the 1/0-toggling, $c_n(t)$ exhibits a non-zero DC component. Hence, the multiplication by $c_n(t)$ can be considered an unbalanced mixing of the DAC signals maintaining the baseband, consisting of the original DAC spectrum $S_{DAC}(f)$, its replica $S_{DAC}(f_{s,d}+f)$ at $f_{s,d}$, and its mirror-image $S_{DAC}(f_{s,d}-f)$. Since the Nyquist frequency of the 2:1 AMUX is at $f_{s,d}$, only the original DAC spectrum and its mirror image contribute to the first Nyquist band of the AMUX output signal. These contributions are shown in the "baseband" and "mirror-image" columns in Figure 2. There, the baseband column displays the original DAC1,2 output spectra with their first two Nyquist bands up to $f_{s,d}$. Due to the unbalanced mixing of the AMUX, these spectra appear directly at the output and are marked by M1,2 according to the mixer

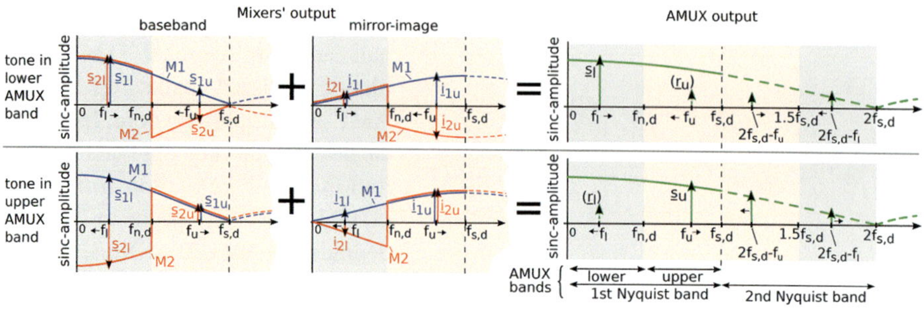

Fig. 2. Spectra at the AMUX-output of a 2:1 AMUX.

of the respective input. Considering zero order hold (ZOH) DAC signals, the DAC output amplitude drops by $\text{sinc}(f/f_{s,d})$: $= \sin(\pi f/f_{s,d})/(\pi f/f_{s,d})$ to sinc ½ ≈ 63.7 % (−4 dB) at the DAC Nyquist frequency $f_{n,d} = f_{s,d}/2$, and to zero at $f_{s,d}$. This behavior is shown by the sinc-envelope of the M1,2 signal amplitude in Figure 2. The second DAC signal is shifted against the first one by $T_{s,a} = T_{s,d}/2$ (i.e. 180°) to align the center of their ZOH signal to the sampling pulses $c_{1,2}(t)$. The 180° phase shift of the second DAC signal leads to a sign change in its second Nyquist band in the M2 spectrum, shown by inversion of the sinc-envelope in Figure 2. The mirror image of the baseband at $f_{s,d}$ is shown in the second column of Figure 2 by its mirrored sinc-envelope. Because of the respective phase shifts of the sampling signals, for those mirror images the same sign considerations hold as in the baseband. Higher order images also contribute to the output signal, but due to their sinc weighting, they are of minor influence.

To show how a tone of a certain frequency is created, the AMUX output spectrum can be subdivided into a lower and an upper AMUX band, as defined in Figure 2, that correspond to the first two Nyquist bands of the DAC signals. Hence, a tone at f_l, f_u generated in the lower/upper AMUX band comes with an alias tone at $f_{s,d} - f_l, f_{s,d} - f_u$ in the upper/lower AMUX band, respectively. In Figure 2, the baseband tones are depicted by phasors $\underline{s}_{1,2l}$ and $\underline{s}_{1,2u}$ for the lower and upper band of M1,2. Phasors pointing up/downwards exhibit a 0°/180° phase shift, respectively. The respective image phasors are represented by phasors $\underline{i}_{1,2l}$ and $\underline{i}_{1,2u}$. The two rows of Figure 2 illustrate tone generation for both lower and upper AMUX bands. For a generation of a tone at f_l in the lower band, the two DACs have to output signals in phase (i.e. in the first Nyquist band of the DACs), while they are of opposite phase for a tone at f_u in the upper band.

The final AMUX output signal constitutes of the superposition of the baseband and mirror image signals in either row of Figure 2. Ideally, the intended tones superimpose constructively, while the alias tones cancel out. The superposition of the baseband and mirrored $\text{sinc}(f/f_{s,d})$ envelopes results in a total $\text{sinc}(f/(2f_{s,d}))$ envelope as shown in the right column of Figure 2. Because of this sinc-envelope and an AMUX output 1st Nyquist bandwidth of twice the DACs' 1st Nyquist bandwidth, the 2:1 AMUX output mimics a Super-DAC with doubled sampling rate.

In the presence of amplitude or timing mismatches of the respective data and sampling signals, residual tones $\underline{r}_{l,u}$ persist in the output signal. Those residual tones originate from imperfect cancellation in amplitude or phase of the alias tones and can be caused by multiple sources in the AMUX setup. For high signal path

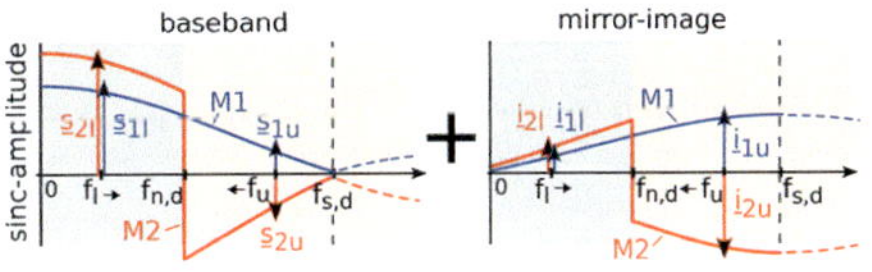

Fig. 3. Example of an amplitude mismatch.

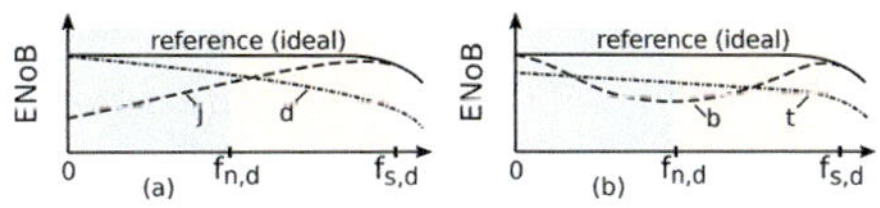

Fig. 4. Basic types of ENoB vs. frequency characteristics of a 2:1 AMUX. (a) Diverged, d-type, and joined, j-type. (b) Tilted and translated, t-, and bathtub, b-type [11].

linearity and high SINAD DAC signals, they become the dominant impairment on AMUX resolution. As an example shown in Figure 3, an amplitude mismatch between M1,2 inputs is considered for lower band tone generation. This mismatch can originate from different DAC amplitudes or a gain mismatch along the AMUX signal paths. In the shown case, M2 exhibits a higher amplitude. For the wanted tone, the higher M2 amplitude increases the baseband $|\underline{s}_{21}|$ as well as the mirror image amplitude $|i_{21}|$ and thus the total output amplitude, which still approximately follows a sinc-envelope. For the alias tone, the M1,2 tones retain their respective 0°/180° phase but exhibit different amplitudes; thus, their sum is nonzero, whereby a residual tone remains. The SINAD used for ENoB calculations depends on the ratio of the wanted signal and the residual tone. Depending on the amplitude progression of wanted and residual tone along the baseband and mirror-image sinc-envelopes, characteristic ENoB vs. frequency traces are created. In the example of the amplitude mismatch in Figure 3 the ratio between wanted and residual tone amplitude increasingly decreases over frequency. This holds especially for tones in the upper AMUX band, where the sinc-envelope for the residual tones reaches its maximum for $f_u = f_{s,d}$. This behavior results in t-type ENoB characteristic [11] shown in Figure 4. To minimize the amplitude mismatch, a symmetrical AMUX circuit design as well as matched DAC signal paths are important. Analogous to the amplitudes, also the signal and clock phases have to exactly match the 0°/180° relation to cancel out perfectly. As shown in [11], any imperfection causes one of the ENoB characteristics in Figure 4, where (d) and (j) are related to clock offset and delay, respectively, whereas (b) relates to imperfections in the clock-to-signal timing, and signal-to-signal timing. The individual characteristics of wanted and residual tones of these characteristics are utilized in [11] to develop tailored calibration routines for the AMUX setup.

To extend the frequency-domain model in Figure 2 to a 4:1 AMUX, the model needs to be adapted to the differences between 2:1 and 4:1 versions of the AMUX. However, for the DAC signals, generally, the same ZOH sinc-envelope properties as for the 2:1 AMUX are considered. In favor of clarity, in the baseband column of the 4:1 AMUX spectra in Figure 5, only two out of four DAC input signals are shown in one diagram. Thereby, the total of four signals are shown in the two diagrams on top of one another. Also, only one example is given for the tone generation in the lowest (first out of four) AMUX band. For the 4:1 AMUX, the baseband spectrum $S_{DAC}(f)$, as well as the mirror-image $S_{DAC}(f_{s,d} - f)$, its replica $S_{DAC}(f_{s,d} + f)$, and the second mirror image $S_{DAC}(2f_{s,d} - f)$ contribute to the first AMUX Nyquist band. As the sample time $T_{s,a} = T_{s,d}/4$ is halved compared to the 2:1 AMUX, also the related phase shifts in the four AMUX bands are halved. While the time shifts and properties of M1,3 are the same as for M1,2

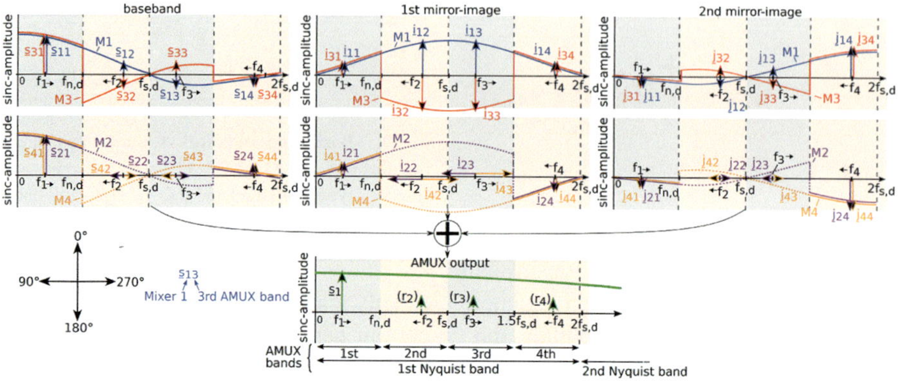

Fig. 5. Composition of the AMUX-output spectrum of a 4:1 AMUX for a tone f_1 in the first AMUX band. The frequencies f_2, f_3, and f_4 are the alias tones in the other AMUX bands. The $\underline{i}$ and $\underline{j}$ represent first and second mirror-image tones, respectively. Dotted lines for M2,4 represent the sinc-envelopes phase-shifted by 90°/270° in the up/downward direction.

of the 2:1 AMUX, for every second input signal (M2,4 in Figure 5), 90° and 270° phase shifts occur in the second and third AMUX bands. Due to the shortened sampling pulses compared to the 2:1 AMUX, the output amplitude follows a sinc($f/(4f_{s,d})$) envelope. In the example in Figure 5, it can be seen that at the wanted tone in the first AMUX band, all four baseband signals superpose constructively, while in the other AMUX bands, the signals cancel out each other. For this example, in the second and third bands, M1,3 cancel out due to respective 0°/180° phase shifts. Also, M2,4 exhibit 90°/270° phase shifts and cancel out, respectively. In the fourth AMUX band, M1,3 have the same phase of 0° and superpose constructively, while M2,4 exhibit 180°, canceling out corresponding M1,3 alias tones. For an imperfect alias tone cancellation, a total of three residual tones can occur for the 4:1 AMUX.

As there are double the inputs and up to three instead of one residual tones compared to the 2:1 AMUX, the calibration of a 4:1 AMUX measurement or simulation is even more complex. Like for the 2:1 AMUX, the origins of wanted and unwanted tones and their sinc-related frequency dependency of their amplitudes are valuable insights for ENoB optimization in the current 4:1 AMUX development.

III. 2:1 AND 4:1 AMUX CIRCUIT CONCEPTS

Multiple concepts to realize the AMUX function are present in literature. These are either based on frequency-domain or time-domain multiplexing [16]. Since the latter is comparatively easy to realize and does not need for any challenging amplitude and phase alignments across the interleaved frequency bands, we focus here on the time-domain approach.

Most time-domain 2:1 AMUX concepts combine two DAC inputs to one output signal, acting like the toggle switch in Figure 1a. The two main toggle switch concepts are based on Gilbert cells and are named clocked-TAS and clocked-SEL, respectively [11]. In the clocked-SEL concept, shown in Figure 6 for the 2:1 AMUX, two linear transadmittance stages (TAS) are biased via constant currents I_0 and driven by DAC signals. The TAS output currents are switched

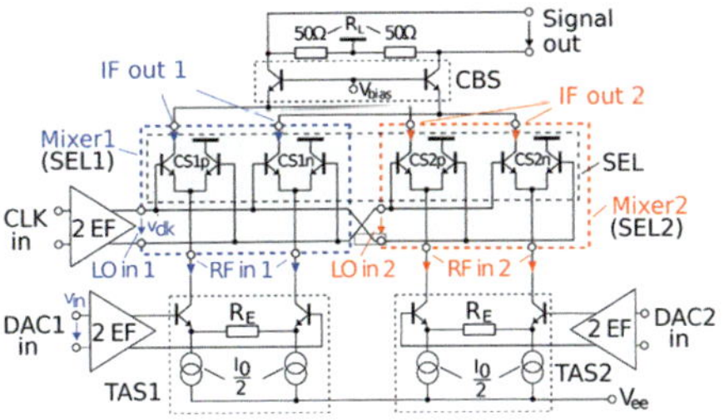

Fig. 6. Circuit principle of the 2:1 AMUX core realized by the clocked-SEL concept.

by selector stages (SEL), which consist of two current switches (CS), each. Those SEL switch the data signals either to the output or to the ground. All inputs incorporate emitter followers, while the output is buffered with a common base stage (CBS). In contrast, the competing clocked-TAS concept (cf. [4]–[6]) switches the TAS bias currents, so that only one of the two TASs is turned on at a time and contributes to the output signal. In [11], it is shown that the effective resolution achievable with this concept is strongly suffering from dynamic currents through the non-linear base-emitter junction capacitance of the TAS transistor. Therefore, in the following, the clocked-SEL is preferred over the clocked-TAS.

The multiplication of the input signal by a sampling signal $c_n(t)$ as introduced in Figure 1b is realized by driving the SEL by a clock voltage v_{clk} at a frequency of $f_{clk} = f_{s,d}$. The sampling signals $c_n(t)$ do not directly represent physical voltages or currents; they rather describe the 1/0-toggling function carried out by the CS/SEL stages. Due to the steep transfer function of a CS, it basically switches its input current at the common emitter node to either of its two outputs depending on the sign of v_{clk}. In case of the 2:1 AMUX, this leads to the data signals being switched alternately and symmetrically with a duty cycle of 1/2 to the output or ground. By inversion of the two clock signals at the respective SEL, only one of the two data signals is present at a time at the AMUX output.

For a higher order AMUX ($N > 2$), also only one data signal at a time is supposed to be present at the output. Thus, the sampling pulses have to be shortened to the AMUX sample time $T_{s,a} = T_{s,d}/N$. This corresponds to a duty cycle of $1/N$ for the sampling signals. Figure 7 proposes a 4:1 AMUX circuit using the clocked-SEL concept. It basically consists of two of the 2:1 AMUX cells of Figure 6, whose output currents are summed up by only one CBS. The SEL2,4 are driven by clock signals shifted by 90°/270° compared to SEL1,3, which need 0°/180° as for the 2:1 AMUX. The lower duty cycle is achieved by the introduction of an offset (OS) to the clock voltage at each SEL (cf. "OS"-input at the driving EF stages). This shortens the time where $v_{clk} > 0$ and thereby reduces the sampling time of the SEL. Assuming sinusoidal clock signals, an offset V_{os} related to the clock amplitude

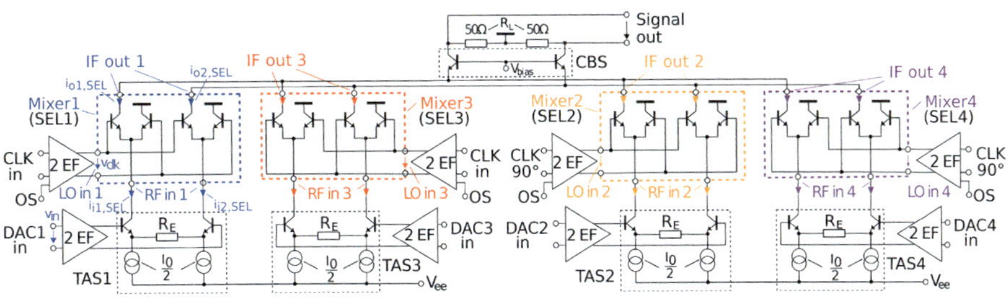

Fig. 7. Circuit principle of the proposed 4:1 AMUX core realized by the clocked-SEL concept.

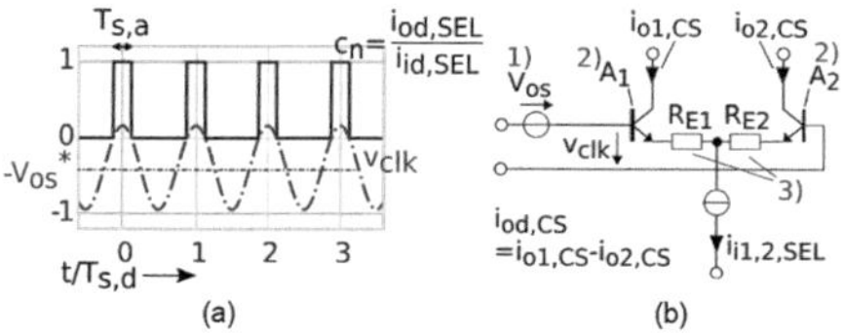

Fig. 8. (a) Clock voltage with offset and resulting idealized sampling signal $c_n(t)$ for a 4:1 AMUX. (b) Three methods to introduce an offset to the transfer characteristic of a CS by (1) offset voltage V_{os}, (2) different transistor areas A_1, A_2, or (3) different emitter degeneration resistors R_{E1}, R_{E2}.

V_{clk} by a factor of $V_{os}/V_{clk} = \cos(\pi/N)$ is needed to achieve a duty cycle of $1/N$. For the case of the 4:1 AMUX shown in Figure 8a, this ratio yields $\sqrt{2}/2 \approx 70.7\%$. It is important to note that this offset depends on the amplitude and the signal shape of the clock signal and is thereby a crucial parameter in design optimization.

In Figure 8b, three different methods to generate the needed offset for a CS are suggested: (1) direct application of an offset voltage V_{os}, (2) transistors with different emitter areas A_1, A_2 and (3) different emitter resistances R_{E1} and R_{E2}. The proposed methods combined contribute to an effective offset voltage of the SEL-CSs

$$V_{OS}^{*} = V_{OS} + V_T \ln\frac{A_2}{A_1} + \frac{R_{E1} - R_{E2}}{2} i_{i1,2,SEL}, \tag{1}$$

where $i_{i1,SEL}$ and $i_{i2,SEL}$ are the TAS output currents at the emitter node input of the respective SEL-CS. If eqn (1) is considered in the transfer characteristic of a SEL-CS, the offset becomes dependent on those actual current values, which result in a non-linear offset and transfer characteristic. In case of only methods (1) and (2) (i.e. $R_{E1} = R_{E2} = 0$) in eqn (1), the single-ended current transfer function

$$\frac{i_{o1,2,SEL}}{i_{i1,2,SEL}} = \frac{1}{2}\left(1 + \tanh\left(\frac{V_{clk}\cos(2\pi f_{clk} t) - V_{OS}^{*}}{2V_T}\right)\right) \tag{2}$$

does not depend on the actual current value, so that the SEL-CS sampling linearly transfers the sample values. Due to the sampling linearity, the single-ended expression in eqn (2) can analogously be expressed for the differential output current of the SEL. However, the size factor A_2/A_1 becomes impractical as due to the logarithm, very large factors (e.g. > 10^5 for V_{clk} = 500 mV) are needed to get usable offsets. Hence, the only reasonable approach to generate the needed offset is to apply offset voltages directly to the clock inputs.

IV. SIMULATION AND MEASUREMENT OF A 2:1 AMUX

The following measurement results are obtained from a 2000 × 900 µm sized AMUX chip realized in IHP SG13G2 technology with f_T = 300 GHz and f_{max} = 500 GHz [21]. The simulated current consumption is 648 mA/35 mA at its two supply voltages of −5.5 V/−2.5 V, respectively. Due to on-chip supply voltage issues, the total measured power consumption is raised by about 30% up to 4.7 W. Nevertheless, the chip shows results good enough to be published, because its ENoB is only marginally lower than the highest reported ENoB in [10] but at a higher sampling rate and with significantly higher clock- and data-path bandwidths (cf. Table I). Furthermore, the AMUX, completely packaged in an RF-module, achieves the highest reported maximum sample rate of 186 GS/s for an SiGe module. Figure 9a shows the layout of the AMUX chip that follows the cell-based design approach introduced in [22]. The high number of pads is used to adjust the currents of the transistor stages and determines the chip size. The clock amplifier chain incorporates three clock buffers and conditions the clock signal for the 2:1 AMUX cell. The AMUX cell

TABLE 1
STATE OF THE ART

Technology (f_T/f_{max})/GHz	Sampling rate	ENoB at ($f_{low} - f_{high}$)***	Bandwidth clock path	Bandwidth signal path	Diff. output voltage swing	Power/supply voltage
[4] 0.5 µm-InP(290/320)	80 GS/s	–	50 GHz	–	–	0.5 W/ −4.5 V
[5] 0.5 µm-InP(290/320)	128 GS/s	–	64 GHz	67 GHz	0.5 V_{pp}	0.54 W/ −4.5 V
[6] 0.25 µm-InP(460/480)	168 GS/s**	–	99 GHz	> 110 GHz	1.5 V_{pp}	0.9 W/ −4.5 V
[18] 0.7 µm-InP(340/410)	100 GS/s	–	–	–	1.1 V_{pp}	–
[19] 0.13 µm-SiGe(300/500)	56 GS/s*	–	60 GHz*	> 67 GHz*	0.75 V_{pp}	1.06 W/ −4.5 V
[14] 0.13 µm-SiGe(470/650)	190 GS/s* **	–	–	> 110 GHz*	1 V_{pp}	1.5 W/–
[20] 55 nm-SiGe(320/370)	100 GS/s	4.9–4.25	≥ 50 GHz	73 GHz	0.4 V_{pp}	0.7 W/ 2.5 V
[10] 55 nm-SiGe(325/375)	120 GS/s	7.7–4.1	64 GHz	50 GHz	0.7 V_{pp}	2.17 W/ −5.7 V
This work 0.13 µm-SiGe(300/500)	186 GS/s****	7.1–3.7	> 70 GHz	> 70 GHz	1.2 V_{pp}	4.7 W/ −5.5 V

* Measured on chip. ** Using pre-filtered input signals. *** f_{low}/f_{high} is the smallest/highest measured frequency for the ENoB. (**** For PAM-2).

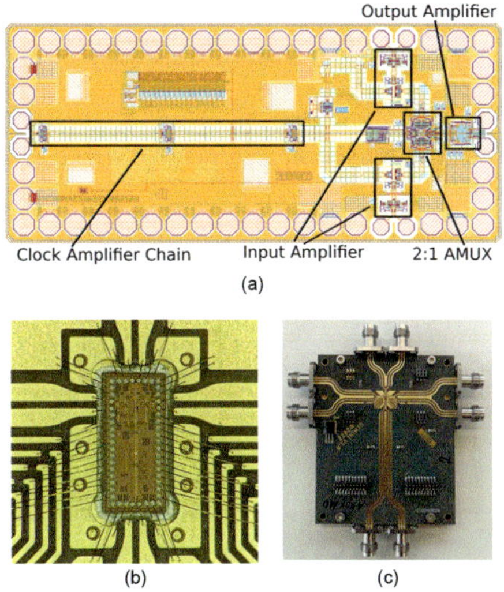

Fig. 9. Chip layout (a), AMUX IC bondwire assembly (b), and RF-module (c).

itself is designed using the clocked-SEL principle as high linearity was the primary concern of the design. Data signals are buffered by input amplifiers, while the AMUX cell is placed in close vicinity to an output amplifier and the chip output, to keep the high-speed signal path short. Figure 9b,c show the chip bondwire and RF module assembly following the principle described in [23]. Target specification for the 2:1 was a maximum sampling rate exceeding 120 GS/s, while maintaining an ENoB of 6 bit. Figure 10a shows on the left the simulated ENoB and amplitude curves for the AMUX at 128 GS/s. On the right hand side, the measured ENoB and amplitude curves are shown. Measurements were carried out using two 8-bit MICRAM DAC5 AWG modules for signal generation, a Keysight E8257D analog signal generator for clock signal generation that drives both AMUX and DAC clocks, and a Keysight N1000A DCA-X sampling oscilloscope, respectively. No digital corrections or data pre-emphasis were applied to any of the measurements. The

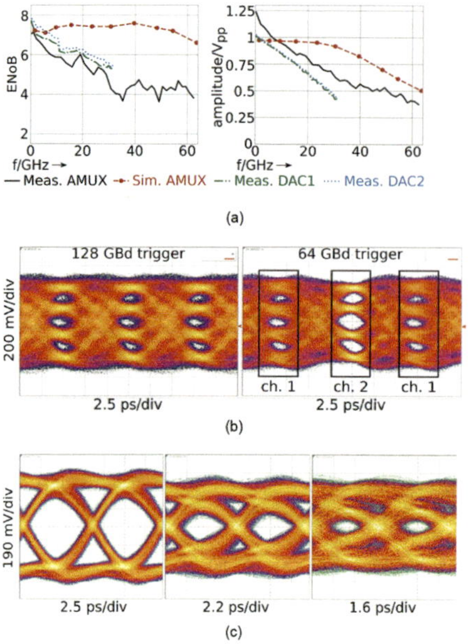

Fig. 10. (a) ENoB and amplitude curves at 128 GS/s. (b) PAM-4 128 GS/s eye diagrams, triggered to 128 GBaud (left) and triggered to 64 GBaud (right). (c) PAM-2 eye diagrams for 120/140/186 GS/s, respectively.

AMUX reaches an ENoB of 7.1 bit for the lowest output frequency, dropping to a minimum of 3.7 bit at 49 GHz. The ENoB curve basically follows the DAC ENoB curve in the lower AMUX band, which is also shown for reference in Figure 10a. It drops relatively fast from 7.1 bit to about 4 bit at a middle output frequency. For higher frequencies, it remains quite constant between 3.7 and 5 bit. Figure 10b shows measured PAM-4 eye diagrams for the AMUX at 128 GS/s. On the left side of Figure 10b, eye diagrams triggered at 128 GS/s are shown. The eye diagrams shown on the right side were triggered at 64 GS/s, showing the two corresponding data channels, alternately. Both channels exhibit different eye openings, which is likely due to a thermal asymmetry in the AMUX chip caused by the increased current consumption. However, the eye opening of channel 2 promises an upward potential if the supply voltage issue was fixed. Finally, Figure 10c shows PAM-2 eye diagrams up to the operating limit of 186 GS/s.

V. CONCLUSION

In this paper, we presented considerations to extend the 2:1 AMUX concept to a 4:1 AMUX. The frequency-domain explanatory model introduced in [11] was extended to the 4:1 AMUX, giving valuable insights for the ENoB calibration of simulation and measurements. Also, for the 4:1 AMUX, a new circuit concept based on the clocked-SEL was shown, providing a new approach for future AMUX designs. Finally, we showed measurements of a 2:1 AMUX module exhibiting a leading maximum ENoB of up to 7.1 bit at 128 GS/s and a maximum sample rate of 186 GS/s.

ACKNOWLEDGMENTS

Part of this work was funded by European Horizon 2020 program TARANTO and co-funded by the German BMBF (grant agreement number 737454), now continued in SHIFT project (grant agreement number 101096256).

REFERENCES

[1] Nokia, "Nokia launches next gen coherent optics to reduce network power consumption by 60 percent," 2023, https://www.nokia.com/about-us/news/releases/2023/02/16/nokia-launches-next-gen-coherent-optics-to-reduce-network-power-consumption-by-60-percent/ [Accessed: May 23, 2024].

[2] Gazettabyte, "The OIF's coherent optics work gets a ZR+ rating," 2024, https://www.gazettabyte.com/home/2024/2/9/the-oifs-coherent-optics-work-gets-

a-zr-rating.html [Accessed: May 23, 2024].

[3] D. Ferenci, M. Grozing, and M. Berroth, "A 25 GHz Analog Multiplexer for a 50GS/s D/A-Conversion System in InP DHBT Technology," in *2011 IEEE Compound Semiconductor Integrated Circuit Symposium (CSICS)*, 2011, pp. 1–4.

[4] M. Nagatani, H. Yamazaki, H. Wakita, H. Nosaka, K. Kurishima, M. Ida, A. Sano, and Y. Miyamoto, "A 50-GHz-bandwidth InP-HBT analog-MUX module for high-symbol-rate optical communications systems," in *2016 IEEE MTT-S International Microwave Symposium (IMS)*, 2016, pp. 1–4.

[5] M. Nagatani, H. Wakita, H. Yamazaki, H. Nosaka, K. Kurishima, M. Ida, and Y. Miyamoto, "A 128-GS/s 63-GHz-bandwidth InP-HBT-based analog-MUX module for ultra-broadband D/A conversion subsystem," in *2017 IEEE MTT-S International Microwave Symposium (IMS)*, 2017, pp. 134–136.

[6] M. Nagatani, H. Wakita, H. Yamazaki, Y. Ogiso, M. Mutoh, M. Ida, F. Hamaoka, M. Nakamura, T. Kobayashi, Y. Miyamoto, and H. Nosaka, "A Beyond-1-Tb/s Coherent Optical Transmitter Front-End Based on 110-GHz-Bandwidth 2:1 Analog Multiplexer in 250-nm InP DHBT," *IEEE Journal of Solid-State Circuits*, vol. 55, no. 9, pp. 2301–2315, 2020.

[7] M. Nagatani, H. Wakita, Y. Ogiso, H. Yamazaki, M. Mutoh, M. Ida, F. Hamaoka, M. Nakamura, T. Kobayashi, Y. Miyamoto, and H. Nosaka, "A 110-GHz-Bandwidth 2:1 AMUX-Driver using 250-nm InP DHBTs for Beyond-1-Tb/s/carrier Optical Transmission Systems," in *2019 IEEE BiCMOS and Compound semiconductor Integrated Circuits and Technology Symposium (BCICTS)*, 2019, pp. 1–4.

[8] M. Nakamura, F. Hamaoka, H. Yamazaki, M. Nagatani, Y. Ogiso, H. Wakita, M. Ida, A. Matsushita, T. Kobayashi, H. Nosaka, and Y. Miyamoto, "1.3-Tbps/carrier net-rate signal transmission with 168-GBaud PDM PS-64QAM using analogue-multiplexer-integrated optical frontend module," in *45th European Conference on Optical Communication (ECOC 2019)*, 2019, pp. 1–3.

[9] R. Hersent, A. Konczykowska, F. Jorge, F. Blache, V. Nodjiadjim, M. Riet, C. Mismer, and J. Renaudier, "160-GSa/s-and-Beyond 108- GHz-Bandwidth Over-2-Vppd Output-Swing 0.5-µm InP DHBT 2:1 AMUX-Driver for Next-Generation Optical Communications," *IEEE Microwave and Wireless Components Letters*, vol. 32, no. 6, pp. 752–755, 2022.

[10] M. Collisi and M. Möller, "A 120 GS/s 2:1 Analog Multiplexer with High Linearity in SiGe-BiCMOS Technology," in *2020 IEEE BiCMOS and Compound Semiconductor Integrated Circuits and Technology Symposium (BCICTS)*, 2020, pp. 1–4.

[11] ——, "Resolution-Related Design Considerations for a 120-GS/s 8-bit 2:1 Analog Multiplexer in SiGe-BiCMOS Technology," *IEEE Journal of Solid-State Circuits*, vol. 56, no. 9, pp. 2624–2634, 2021.

[12] Q. Hu, K. Schuh, M. Collisi, V. Aref, H. Hettrich, R. Schmid, F. Buchali, and M. Möller, "120 GSa/s BiCMOS Analog Multiplexer Enabling 360 Gbit/s DSCM-PCS-256QAM IM/DD Transmission," in *2021 Optical Fiber Communications Conference and Exhibition (OFC)*, 2021, pp. 1–3.

[13] Q. Hu, K. Schuh, V. Aref, F. Buchali, T. Tannert, M. Grözing, M. Berroth, C. Schmidt, J. Schostak, C. Caspar, J. H. Choi, V. Jung- nickel, and R. Freund, "128 GSa/s SiGe Analog Multiplexer Enabling PDM-DSCM-PCS-256QAM Transmission over 80 km SSMF," in *Optica Advanced Photonics Congress 2022*. Optica Publishing Group, 2022, p. SpTh1I.6.

[14] J. Schostak, T. Tannert, M. Grözing, V. Jungnickel, C. Schmidt, H. Rücker, M. Berroth, and R. Freund, "190 GBd PAM-4 Signal Generation using Analog Multiplexer IC with On-Chip Clock

Multiplier," in *2023 18th European Microwave Integrated Circuits Conference (EuMIC)*, 2023, pp. 1–4.

[15] Q. Hu, T. Tannert, M. Grözing, G. Raybon, R. Borkowski, F. Buchali, X. Chen, P. Iannone, G. Rademacher, and R. Ryf, "467 gbit/s net bitrate im/dd transmission using 176 gbd pam-8 enabled by sige amux with excellent linearity," in *2024 Optical Fiber Communications Conference and Exhibition (OFC)*, 2024, pp. 1–3.

[16] C. Schmidt, *Interleaving concepts for digital-to-analog converters.* Springer Vieweg, Wiesbaden, 2020.

[17] H. Yamazaki, M. Nakamura, T. Kobayashi, M. Nagatani, H. Wakita, Y. Ogiso, H. Nosaka, T. Hashimoto, and Y. Miyamoto, "Net-400-Gbps PS-PAM transmission using integrated AMUX-MZM," *Optics Express*, vol. 27, p. 25544, 09 2019.

[18] R. Hersent, A. Konczykowska, F. Jorge, M. Riet, C. Mismer, V. Nodjiadjim, B. Duval, and J.-Y. Dupuy, "Analog-Multiplexer (AMUX) circuit realized in InP DHBT technology for high order electrical modulation formats (PAM-4, PAM-8)," in *2020 23rd International Microwave and Radar Conference (MIKON)*, 2020, pp. 222–224.

[19] T. Tannert, X.-Q. Du, D. Widmann, M. Grözing, M. Berroth, C. Schmidt, C. Caspar, J. H. Choi, V. Jungnickel, and R. Freund, "A SiGe-HBT 2:1 analog multiplexer with more than 67 GHz bandwidth," in *2017 IEEE Bipolar/BiCMOS Circuits and Technology Meeting (BCTM)*, 2017, pp. 146–149.

[20] H. Ramon, M. Verplaetse, M. Vanhoecke, H. Li, J. Bauwelinck, P. Ossieur, X. Yin, and G. Torfs, "A 100-GS/s Four-to-One Analog Time Interleaver in 55-nm SiGe BiCMOS," *IEEE Journal of Solid-State Circuits*, vol. 56, no. 8, pp. 2539–2549, 2021.

[21] M. Inac, A. Fatemi, F. Korndörfer, H. Rücker, F. Gerfers, and A. Malignaggi, "Performance Comparison of Broadband Traveling Wave Amplifiers in 130-nm SiGe:C SG13G2 and SG13G3 BiCMOS Technologies," *IEEE Microwave and Wireless Components Letters*, vol. 31, no. 6, pp. 744–747, 2021.

[22] M. Möller, "Challenges in the Cell-Based Design of Very-High-Speed SiGe-Bipolar ICs at 100 Gb/s," *Solid-State Circuits, IEEE Journal of*, vol. 43, pp. 1877 – 1888, 10 2008.

[23] H.-M. Rein and M. Möller, "Design considerations for very-high-speed Si-bipolar IC's operating up to 50 Gb/s," *Solid-State Circuits, IEEE Journal of*, vol. 31, pp. 1076–1090, 1996.

Oliver Hauck was born in St. Ingbert, Germany, in 1998. He received the M.Sc. degree in systems engineering from Saarland University, Saarbrücken, Germany, in 2023, where he is currently pursuing the Ph.D. degree in electrical engineering. Since 2023, he has been a research assistant with the Chair of Electronics and Circuits, Saarland University. His current research interest includes the optimization methods of (broadband) high-speed BiCMOS circuits with special regard to analog multiplexing.

Fred Buchali received the diploma degree in electrical engineering in 1988 and the PhD degree in 1991 from Humboldt-University Berlin. In 1999, he joined the optical transmission systems group of Alcatel SEL Research and Innovation, where he contributed to optical and electrical impairment mitigation techniques at 10 and 40 Gb/s by analog electrical and optical techniques. During the emergence of coherent transmission technology, he substantially contributed to the investigation of OFDM in optics. His current research interests cover coherent single-carrier and multicarrier systems beyond 1 Tb/s using higher order modulation formats including probabilistic shaping. He authored or co-authored more than 100 journal and conference papers and holds more than 30 patent applications.

Michael Collisi was born in Dudweiler, Germany, in 1990. He received the M.Sc. and Dr.-Ing. degrees in mechatronics from Saarland University, Saarbrücken, Germany, in 2017 and 2024, respectively. From 2017 to 2023, he worked as a research assistant with the Chair of Electronics and Circuits (Saarland University) where he focused on the optimization methods of (broadband) high-speed BiCMOS circuits with special regard to analog multiplexing. In 2023, he joined Keysight Technologies as an IC design engineer in Bochum, Germany.

Michael Möller (M'06) was born in Bebra, Germany, in 1965. He received the Dipl.-Ing. degree in electrical engineering from FH Giessen, Giessen, Germany, in 1987, and the Dipl.-Ing. and Dr.-Ing. degrees in electrical engineering from Ruhr-University Bochum, Bochum, Germany, in 1991 and 1999, respectively. At Ruhr-University Bochum, he was with Arbeitsgruppe Halbleiterbauelemente (Prof. H.-M. Rein), where he focused on the development of high-speed circuits in Si-bipolar technology as key components in fiber-optical transmission systems. Many of his chip designs, which he developed with team members and partners, have achieved world records for speed and performance. In 1999, he joined Micram Microelectronic GmbH, Bochum, Germany and was named 2002 as an Executive Board Member. In 2005, he became a Professor at Saarland University where he is teaching high-speed electronics and circuits. In his main fields of research, he is working on the exploration of performance limits of integrated high-speed circuits and on the development of circuit-design and optimization methods for performance enhancement at high-operating speeds.

Challenges for 2.5D and 3D Integration of InP HBT Technology

Bertrand Ardouin, Tom K. Johansen, Antoine Chauvet, Romain Hersent, Virginie Nodjiadjim, Agnieszka Konczykowska, Nil Davy, Muriel Riet and Colin Mismer

Abstract—**Indium phosphide (InP) double heterojunction bipolar transistor (D-HBT) technology can be co-integrated with silicon (e.g., silicon-on-insulator (SOI), fully depleted SOI (FDSOI), silicon germanium bipolar complementary metal-oxyde semiconductor (SiGe BiCMOS)) and antennas in order to benefit from their superior high-frequency performances in a cost-effective manner (using 2.5D, 3D integration techniques), while benefiting from the higher integration level provided by silicon ICs. This paper presents the strategic and practical InP D-HBT technology development challenges for the successful 2.5D/3D integration of millimeter wave and sub-THz applications from the perspective of InP manufacturing.**

Index Terms—**Double heterojunction bipolar transistor (D-HBT), indium phosphide, InP/InGaAs, modeling, terahertz (THz), 2.5D, 3D integration.**

This work received funding from Chips Joint Undertaking (Chips JU) under grant agreement number 101096256 (project SHIFT) and grant agreement number 101139842 (project Move2THz). The Chips JU receives support from the European Union's Horizon Europe research and innovation program and the National Authorities.

Bertrand Ardouin (bertrand.ardouin@3-5lab.fr), Antoine Chauvet (antoine.chauvet@3-5lab.fr), Romain Hersent (romain.hersent@3-5lab.fr), Virginie Nodjiadjim (virginie.nodjiadjim@3-5lab.fr), Agnieszka Konczykowska (agnieszka.konczykowska@3-5lab.fr), Nil Davy (nil.davy@3-5lab.fr), Muriel Riet (muriel.riet@3-5lab.fr), and Colin Mismer (colin.mismer@3-5lab.fr) are with III-V Lab, F-91767 Palaiseau, France. Tom K. Johansen (tkj@space.dtu.dk) is with DTU, 2800 Kgs. Lyngby, Denmark.

I. INTRODUCTION

In order to transfer the ever-increasing amount of data stored and processed in the digital world from one point to another, analogue transmission channels are unavoidable and are becoming today (ironically) the real bottleneck of our information society. Future 6G communications, Cloud computing and Artificial Intelligence (AI) altogether, to cite only a few applications, are expected to consume more transmission capacity (see Figure 1) than what can be provided by the optical fiber and wireless communications systems' projected capacity scaling. In order to anticipate this huge demand of bandwidth and higher operating frequencies (and consequently the need for very high-speed electronics), researchers have been intensively developing high-speed silicon technologies. It is well known that the maximum oscillation frequency (f_{MAX}) of silicon CMOS (and its various silicon-on-insulator (SOI) and fin field-effect transistors (FinFET) variants) does not improve anymore with gate length reduction since the 28 nm node [1] [2], and is consequently not the technology of choice for future millimeter wave (mmWave) and sub-THz transceiver front-end, despite their superior integration level capabilities and lower cost. As an alternative, SiGe BiCMOS technologies have experienced a remarkable

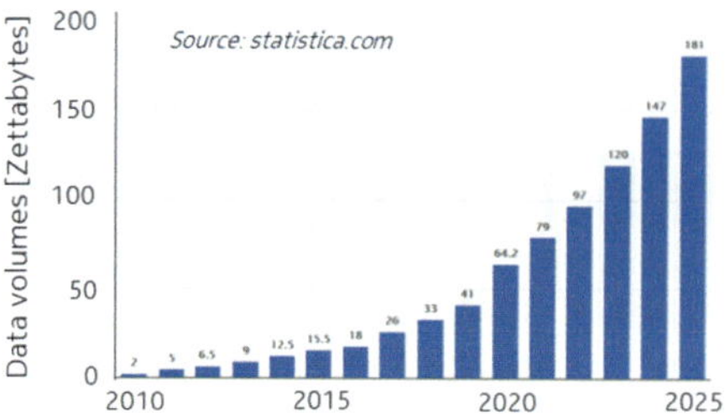

Fig. 1. Volume of data/information created, captured, copied, and consumed worldwide.

performance increase over the last two decades [3][4], making them the technology of choice for radio frequency (RF) front-ends, whenever pure CMOS technologies cannot meet the specifications (e.g., for high-frequency (HF) operation). Nevertheless, for next-generation communication and sensing applications targeting the 100–300 GHz bands, for which the use of antenna arrays and beam steering is necessary to compensate for atmospheric attenuation, the limits of SiGe BiCMOS technology are perceived. Actually, power consumption is a critical metric for large power amplifier (PA) – antenna arrays, making the power added efficiency (PAE) specifications of the PA out of reach of present SiGe BiCMOS processes. Therefore, more exotic technologies, such as indium phosphide (InP) heterojunction bipolar transistor (HBT) technologies, are presently considered as a potential contender. Interestingly, research is gaining traction in this area lately, and literature dealing with CMOS or SiGe BiCMOS hetero-integration (or 2.5D/3D packaging) with InP is flourishing. Indeed, from a pure material performance point of view, there is no debate that InP is superior to silicon in terms of speed, but significant challenges have to be overcome to make it a viable option for future mmWave and sub-THz applications. This paper presents these challenges from a technical, practical, and economical perspective. Section II will present the state-of-the-art and will emphasize InP HBT performance advantages over other high-speed technologies. Section III will detail the challenges for InP high-volume manufacturing, while Section IV will present the challenges related to 2.5D/3D integration. Section V will introduce the requirements regarding advanced compact modeling of InP HBTs, and finally a conclusion will be drawn in Section VI.

II. TECHNOLOGY STATE-OF-THE-ART AND COMPARISON

Figure 2 presents a summarized state-of-the-art of semiconductor technologies suitable for mmWave and sub-THz applications (gallium nitride (GaN) technologies are intentionally not shown as they are commonly considered as more suitable for applications below e.g. 100 GHz and higher power). Figure 2 presents the geometrical mean of f_T and f_{MAX} on the y-axis (as this figure of merit (FoM) is a balanced indicator of transistor's high frequency performance) versus breakdown voltage (or supply voltage for CMOS). As it can be observed, the various technology trends follow roughly the "pseudo" Johnson limit (we use the term "pseudo" here, as the standard Johnson limit refers to f_T only): the higher the transistor speed is, the lower is its voltage drive capability. Note that some moderate deviations from the hyperbolic trend are noticeable in Figure 2, which is due to the positive impact of smaller lithography on f_{MAX} via external parasitic reduction. This, however, does not dramatically change the conclusions that can be drawn from this graph. When looking only at the y-axis, we can note first that the CMOS limitations mentioned in the introduction are clearly visible. Also, it is observed that the most advanced SiGe HBTs (with f_{MAX} > 600 GHz, e.g., [3]) are able to catch up with a majority of the InP HBT technologies. This observation is certainly one of

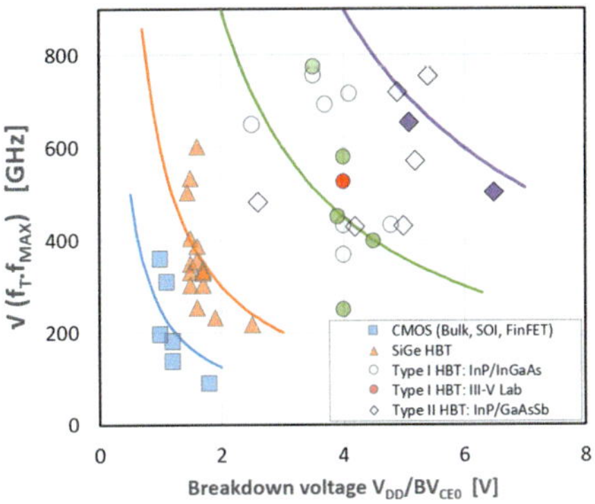

Fig. 2. Geometrical mean of f_T times f_{MAX} versus breakdown voltage. Filled markers represent technologies with known monolithic microwave integrated circuit (MMIC) fabrication capabilities, and empty markers represent technologies able to demonstrate transistors only (no known published circuit results).

the reasons why some "silicon exclusive advocates" argue that InP is not needed in the semiconductor ecosystem apart for some extremely reduced niche markets (astronomy observation, high-end characterization equipment, etc.).

This assertion needs to be reconsidered somehow, when considering the x-axis (breakdown voltage): while SiGe HBTs can almost play on par with InP HBTs in case of small signal operation, this is a misleading conclusion when considering large signal operation (PAs, for instance, which are a key element of future 6G transmission systems).

Actually, the breakdown voltage of Type I HBTs is more than two times larger than that of SiGe HBTs, and the difference is even more pronounced for Type II HBTs which exhibit impressive $\sqrt{(f_T \cdot f_{MAX})} \cdot BV_{CEO}$ product > 3600 GHz.V [5][6][7]. This brings a significant advantage in terms of power drive capability (P_{SAT}) and in terms of PAE for mmWave PA design (noting that the device efficiency is proportional to $1-V_{knee}/V_{BV}$, where V_{knee} is the knee voltage and V_{BV} is the breakdown voltage of the device).

Another important information that can be extrapolated from Figure 2 is the perspective of future improvements. Considering that independently of the technology, optimizing the transistor for higher speed (i.e. higher f_T and f_{MAX}) results in trading off the breakdown voltage (i.e., moving along the hyperbola imposed by Johnson limit), InP HBTs have much more room for improvement. Indeed, going "beyond" the Johnson limit for a given material system requires major structural or technological modifications, which usually come at the expense of process complexity or other performance penalties. To summarize, not only InP HBT technologies currently have a performance advantage (especially when considering PAs above 100 GHz), but they also have the highest potential for (straight forward) improvement in the future.

III. CHALLENGES TOWARDS HIGH VOLUME INP HBT MANUFACTURING

Although InP HBT technologies offer unparalleled performance advantages, its market adoption has been limited to only a few application areas and very limited volume due to some major manufacturing limitations with respect to silicon: higher process cost, lower manufacturability, material brittleness and scarcity, smaller wafer size, and incompatible back-end with mainstream silicon packaging solutions.

A. Fabrication Cost and Other Considerations

InP HBT processes are typically limited to small diameter wafers: many academic labs still use 2″ wafers, while industrial labs and foundries are typically relying on 3″, 4″, or (rarely) 6″ wafers. This is an obvious limitation to move to

Fig. 3. InPoSi wafer fabrication principle (SOITEC SMARCUT™ process).

high-volume manufacturing and to benefit from the economy of scale prevailing in the silicon world. Therefore, research is ongoing to develop InP HBT processes on InPoSi (InP on silicon) wafers (e.g., MOVE2THZ project [8]). InPoSi is a fabrication process consisting in bonding a thin slice of InP from an InP donor wafer on a silicon wafer (see Figure 3). The donor wafer can be reclaimed several times (which is a mitigation to indium material scarcity) and the resulting InPoSi wafer is therefore mechanically more robust, thus improving manufacturability. Moreover, InPoSi can be used to produce 8″ or 12″ wafers through "wafer tiling," thus solving the (donor) wafer size issue mentioned earlier and reducing raw InP wafer cost. Despite this improvement on raw material, the initial epitaxy step still constitutes a major cost contributor that cannot be fully mitigated by the wafer size increase. Therefore, in order to circumvent this problem, alternative paths for cost reduction need to be employed.

B. Alternative Strategies Towards High-Volume Manufacturing

One of the simplest options for reducing InP die cost is to reduce the chip footprint (i.e., minimizing the cost for a given functionality). The solution to this problem is not obvious because analogue and RF functions do not scale in a straightforward manner (as opposed to digital functions that scale quadratically with technology minimum feature size).

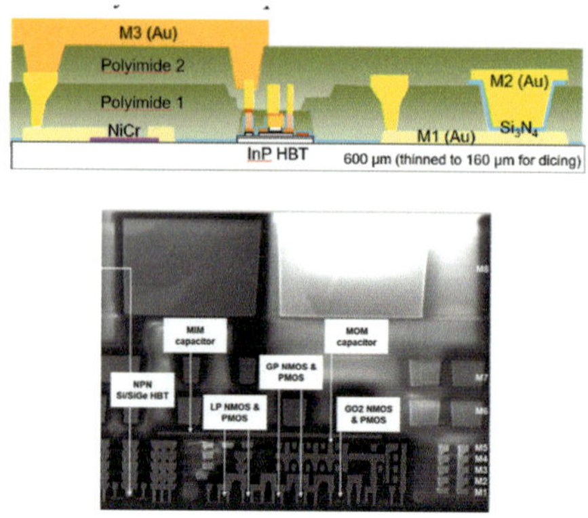

Fig. 4. BEOL of line of the InP HBT technology from III-V Lab (top, not to scale) and BiCMOS55 from ST microelectronics [9] (bottom).

Typical InP HBT technologies have 2–4 relatively thin metal layers. More advanced back-end of lines (BEOLs; i.e., more metal layers, thicker metals, and inter-metal dielectric layers) can significantly reduce both dielectric losses at mmWave and/or the on-chip matching networks' footprint and power combining stages (see Figure 4 for an example of comparison of a typical SiGe BiCMOS [9] and of a typical InP HBT BEOL [10]). Although most of the (dense routing) lower metal layers of silicon technologies may be of little use in an optimal InP HBT process, the top thicker metal layers could be advantageously used for mmWave interconnections. In parallel, development work for higher density routing also needs to be conducted, as most InP HBT manufacturers sometimes use (unnecessary) conservative design rules (lift-off techniques are widely used in InP HBT processes, which partly explain these stringent design rules, but part of the layout restrictions simply come from more simplistic design rules implementation in computer-aided design (CAD) verification tools). Another issue limiting die size reduction is the integrated capacitors and resistors that consume a lot of space on the chip. Compared to the wide variety of integrated RF resistors available in silicon technologies (ranging from a few Ω/□ to several kΩ/□) and double density integrated MiM capacitors (typically from 2 to 4 fF/μm²), the typical single NiCr resistor layers (~50 Ω/□) and Si_3N_4 MiM capacitors (typically below 0.5 fF/μm²) available in typical InP HBT technologies are another limiting factor. Finally, through substrate vias (TSVs) typically used to improve the IC HF stability and eliminate parasitic substrate modes, when available in a given process, consume a large footprint: finding alternatives (such as the use of thin film microstrip lines (TFMSLs)) would be beneficial for the overall die size reduction.

Ultimately, InP area reduction could be pushed even further by reducing the InP die size to the transistor area and its connections pads [11]. This is one of the possible options allowed by co-integration: reducing the die size to its minimal requirement, that is, down to a single transistor. One option is to replace an MMIC (e.g., a Power Amplifier (PA) or Low Noise Amplifier (LNA), including its connections, matching and stabilization input and output networks, decoupling capacitances and its biasing transmission lines and resistors) by a single multi-finger transistor. Considering the example of the PA design shown in Figure 5 (left), whose size is 1.2 × 1.5 mm² and a typical InP HBT structure of 260 × 160 μm² shown in Figure 5 (right), the area reduction ratio can be as high as 40 (and so the cost of the InP die in the bill of material). This option implies other challenges that still need to be solved: all the removed functions need to be efficiently realized in CMOS or SiGe BiCMOS and connected via low loss interconnections. Moreover, in order

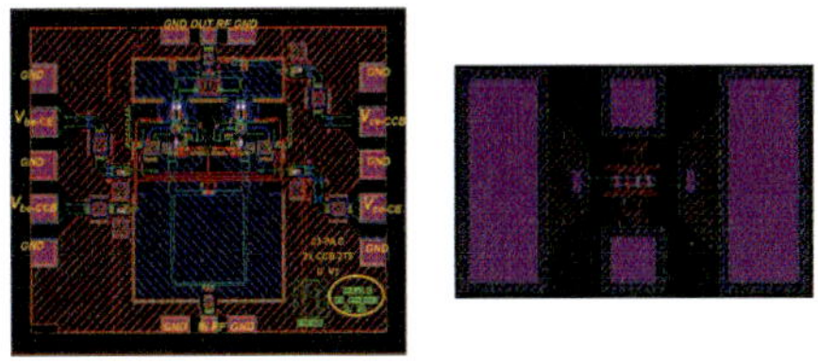

Fig. 5. PA MMIC chip design 1.2 × 1.5 mm² (III-V Lab) (left), multi-finger transistor structures in its RF pads 260 × 160 μm² (right).

to efficiently co-design the silicon chip with the InP HBT transistor, significant effort has to be made in the compact modeling of InP HBTs and in the electrical–electromagnetic co-simulation of the whole system in a complex multi-material environment.

IV. CHALLENGES RELATED TO 2.5D/3D INTEGRATION

A. *Generic Requirement for Hetero-integration*

Describing all possible 2.5D/3D hetero-integration schemes of InP HBT technology with silicon dies is beyond the scope of this paper; it is still interesting to review the requirements on the InP technology side in terms of BEOL. Figure 6 shows some possibilities for SiGe/CMOS co-integration with InP. While SiGe die flip chipping is widely used, flip chipping InP dies is not straightforward, especially for PAs due to thermal constraints. The gold back-end and lack of top passivation (making InP BEOL incompatible with copper pillars), heat dissipation, and HF de-tuning are among the difficulties to overcome. Regarding thermal aspects, the use of InPoSi wafers is still unclear and will remain unclear until the first HBTs become available on InPoSi (while silicon thermal conductivity is better than InP, thermal barriers between InP and Si can mitigate thermal dissipation gains.

Fig. 6. 2.5D/3D integration schemes examples (SiGe in blue and InP in brown).

Moreover, significant work has to be done on the back-end to (i) make it compatible with mainstream flip-chip techniques (finishing/passivation) and (ii) make the die more immune to de-tuning. For this, the InP chip BEOL should have an optimized back-end (more metal layers and thicker metals) to allow low-loss TFMSLs with potentially top and bottom ground planes in order to confine the electromagnetic (EM) field within the back-end. The ultimate goal in this matter is to bring InP dies close to the "chiplet" concept.

Embedding (see Figure 6) is the solution that has been chosen within the SHIFT project [12] for InP/BiCMOS co-integration demonstration. This solution in [13] allows an interesting design option to co-integrate the SiGe BiCMOS chip, the InP PA, and the antenna in order to demonstrate beam-steering capabilities. Demonstration of this principle is currently ongoing as part of the SHIFT project activities. The challenges on InP technologies are the compatibility of BEOL with laser vias (metal thickness), BEOL finishing, and lack of backside metallization and TSV for the considered InP HBT technology (for heat dissipation).

B. Specific Technology Requirements for mmWave Applications and Hetero-integration with Silicon

Many InP HBT processes include TSV and backside metallization, which is currently not the case for III-V Lab's InP D-HBT technology. The III-V Lab technology [14] is natively optimized for high-speed analogue–digital electro-optical interfaces (for optical fiber applications): transistor's figures of merit are tuned for fast current switching, high voltage drive, wide band operation, and low process steps count. Due to the fast switching nature of the main application of this technology, TSV and backside metallization are therefore not considered as essential, taking into account the added process complexity and fabrication time. As shown in Section III.C, TSVs are also questionable regarding chip space requirements. TSV and backside metallization are required for mostly two aspects: (i) microstrip lines design and (ii) parasitic modes suppression. Regarding (i), as application frequency increases, the feasibility of microstrip lines becomes more and more difficult, since the substrate needs to be thinned to, e.g., values below 100 or even 50 μm to remain within the useful line impedance range, thus making the chips very fragile and degrading the process manufacturability. With this respect, switching to an alternative, such as TFMSL, is certainly desirable.

Regarding (ii), let us consider an example of parasitic mode excitation in a test structure based on coplanar wave transmission lines (see EM simulations results at 149 GHz shown in Figure 7, corresponding to a transmission dip of about

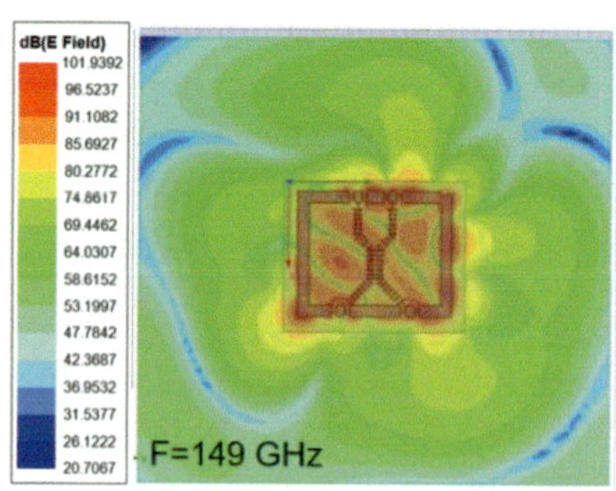

Fig. 7. EM simulation of a differential thru in coplanar wave transmission lines (corresponding to a circuit access). Excited parasitic modes are visible at 149 GHz.

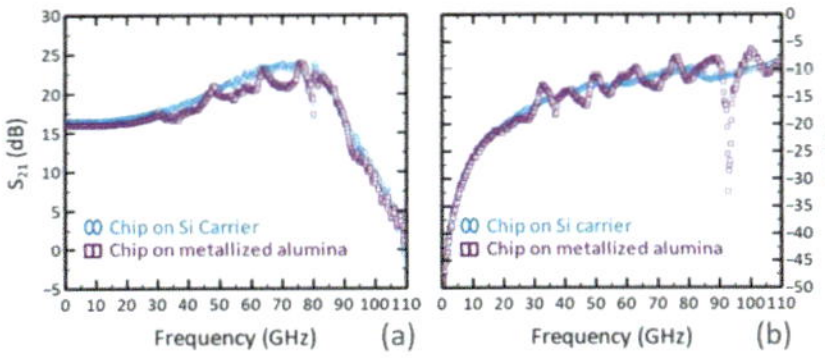

Fig. 8. Measured S parameters of an InP D-HBT AMUX-driver IC. S_{11} (left) and S_{21} (right) versus frequency, IC on a metallized ceramic carrier (red), and IC on a silicon carrier (blue).

8 dB). These parasitic modes' excitations are very detrimental for PA and LNA performances in, e.g., D- and G-bands, as they create dips and resonances in the IC response (to a certain extent, this is less detrimental for fast-switching digital circuits). Figure 8 shows an example of S-parameter measurements of an AMUX-driver IC [15] on both a gold-plated ceramic carrier and a silicon carrier. Indeed, parasitic modes excitation is strongly reduced on silicon, suggesting that the lossy silicon acts as an absorber. Therefore, future InP HBT technologies on InPoSi substrate may be less sensitive to parasitic modes excitations, which would remove the need for TSVs.

V. COMPACT MODELING CHALLENGES

Compact (SPICE-like) modeling of InP HBT technologies definitely lacks beyond silicon. Historically, due to missing suitable compact models for III-V HBTs, the III-V community mostly relied on hand-crafted and/or hand-customized models using either behavioral sources or symbolically defined languages. These models, although sometimes fairly elaborate and accurate for a specific application and/or bias and temperature range of operation [9], usually do not satisfy a wide variety of design needs, like e.g., BiCMOS models and process design kits (PDKs) do. Physics-based, scalable,

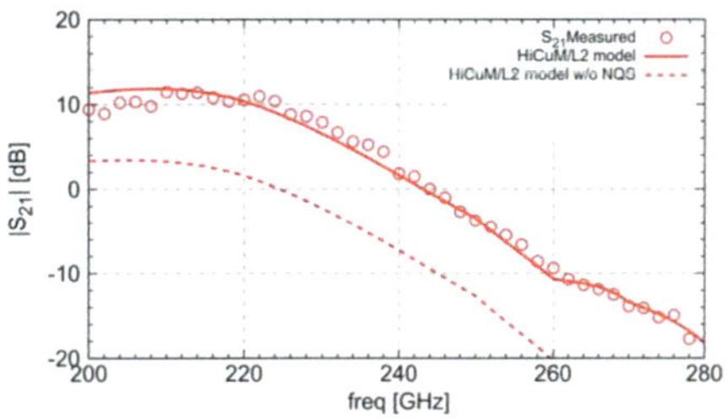

Fig. 9. Measurements/simulation comparison of the S_{21} parameter for a 220 GHz SiGe BiCMOS integrated LNA. Difference between simulations using HiCuM/L2 model with (solid line) and without (broken line) NQS effects.

statistical, and predictive models [14][16][17] are highly desirable to provide designers with larger search space and flexibility in their design points explorations, especially for mmWave and sub-THz applications where each and every dB of gain matters. Moreover, standardized models such as HiCuM/L2 [18][19] are available on virtually every CAD proprietary or open-source software as their Verilog-A code is available to users. An illustration of the need for advanced physics-based compact models such as HiCuM is the often underestimated effect for, e.g., 140–300 GHz frequency range, of non-quasi static (NQS) effects on HF linear gain prediction [20]. As depicted in Figure 9, neglecting NQS effects (which become significant when the application frequency approaches approximately $f_T/3$) can lead to severe discrepancies in the prediction of linear gain in amplifying stages.

VI. CONCLUSION

Today, InP HBT technology is confined to niche markets, thus limiting investment towards high-volume manufacturing and in turn limiting the usage of the technology for medium- to large-scale markets. This cycle can be broken with strategic investments in the InP ecosystem, consisting only in fraction of what is necessary for silicon process developments. While InP HBT will remain a low integration level technology, it can definitely benefit to demanding mmWave applications as a complementary toolset of well-established silicon technologies via hetero-integration, as well as for >Tb/s optical communications. InP HBT cannot and will not replace SiGe BiCMOS, but can adequately complement their already powerful capabilities, provided adequate and ambitious (yet not out of reach) developments are achieved.

ACKNOWLEDGMENT

The authors would like to thank C. Bolognesi for discussions about Type II InP HBT state-of-the-art, F. Brunier for discussions about InPoSi fabrication process, P. Chevalier for sharing the B55 BEOL picture, and F. Gianesello for discussions about hetero-integration.

REFERENCES

[1] S. P. Voinigescu et al., "A Study of SiGe HBT Signal Sources in the 220–330-GHz Range," in IEEE Journal of Solid-State Circuits, vol. 48, no. 9, pp. 2011–2021, Sept. 2013, doi: 10.1109/JSSC.2013.2265494.

[2] H.-J. Lee et al., "Intel 22nm FinFET (22FFL) Process Technology for RF and mm Wave Applications and Circuit Design Optimization for FinFET Technology," 2018 IEEE International Electron Devices Meeting (IEDM), San Francisco, CA, USA, 2018, pp. 14.1.1–14.1.4, doi: 10.1109/IEDM.2018.8614490.

[3] H. Rücker and B. Heinemann, "Device Architectures for High-speed SiGe HBTs," 2019 IEEE BiCMOS and Compound semiconductor Integrated Circuits and Technology Symposium (BCICTS), Nashville, TN, USA, 2019, pp. 1–7, doi: 10.1109/BCICTS45179.2019.8972757

[4] P. Chevalier et al., "SiGe BiCMOS Current Status and Future Trends in Europe," 2018 IEEE BiCMOS and Compound Semiconductor Integrated Circuits and Technology Symposium (BCICTS), San Diego, CA, USA, 2018, pp. 64–71, doi: 10.1109/BCICTS.2018.8550963.

[5] A. M. Arabhavi et al., "THz InP/GaAsSb DHBTs with Record fAVG=800 GHz: Characterization to 330 GHz," 2023 International Electron Devices Meeting (IEDM), San Francisco, CA, USA, 2023, pp. 1–4, doi: 10.1109/IEDM45741.2023.10413850.

[6] S. Hamzeloui et al., "Multi-Finger 250-nm InP/GaAsSb DHBTs with Record 37.3 % Class-A PAE at 94 GHz," 2023 IEEE BiCMOS and Compound Semiconductor Integrated Circuits and Technology Symposium (BCICTS), Monterey, CA, USA, 2023, pp. 145–148, doi: 10.1109/BCICTS54660.2023.10310957.

[7] A. M. Arabhavi et al., "InP/GaAsSb Double Heterojunction Bipolar Transistor Emitter-Fin Technology With fMAX = 1.2 THz," in IEEE Transactions on Electron Devices, vol. 69, no. 4, pp. 2122–2129, April 2022, doi: 10.1109/TED.2021.3138379.

[8] KDT-JU Project MOVE2THZ number: 101139842 (HORIZON-KDT-JU-2023-2-RIA-Topic-1). Web site not available at the time of publication.

[9] P. Chevalier et al., "A 55 nm triple gate oxide 9 metal layers SiGe BiCMOS technology featuring 320 GHz fT / 370 GHz fMAX HBT and high-Q millimeter-wave passives," 2014 IEEE International Electron Devices Meeting, San Francisco, CA, USA, 2014, pp. 3.9.1–3.9.3, doi: 10.1109/IEDM.2014.7046978.

[10] R. Hersent et al., "InP DHBT Linear Modulator Driver With a 3-Vppd PAM-4 Output Swing at 90 GBaud: From Enhanced Transistor Modeling to Integrated Circuit Design," in IEEE

Transactions on Microwave Theory and Techniques, vol. 72, no. 3, pp. 1618–1633, March 2024, doi: 10.1109/TMTT.2023.3305150.

[11] *F. Gianesello, private communication 2023*

[12] *KDT-JU SHIFT project, https://www.shiftkdt.eu/*

[13] I. Peppas, H. Takahashi, J. Yip, E. Schlaffer, H. Paulitsch and W. Bösch, "Embedding of High Power RF Transistor Dies in PCB Laminate," *2022 52nd European Microwave Conference (EuMC)*, Milan, Italy, 2022, pp. 444–447, doi: 10.23919/EuMC54642.2022.9924281.

[14] N. Davy *et al.*, "InP DHBT Analytical Modeling: Toward THz Transistors," in *IEEE Transactions on Computer-Aided Design of Integrated Circuits and Systems*, vol. 42, no. 11, pp. 4102–4111, Nov. 2023, doi: 10.1109/TCAD.2023.3257706.

[15] R. Hersent *et al.*, "100 GBaud DSP-free PAM-4 Optical Signal Generation Using an InP-DHBT AMUX-driver and a Thin-Film Lithium Niobate Modulator Assembly" *2022 IEEE BCICT Symp.,* Monterey , CA, USA, pp. 149–152, doi: 10.1109/BCICTS54660.2023.10310947.

[16] C. Mukherjee *et al.*, "Scalable Compact Modeling of III–V DHBTs: Prospective Figures of Merit Toward Terahertz Operation," in *IEEE Transactions on Electron Devices*, vol. 65, no. 12, pp. 5357–5364, Dec. 2018, doi: 10.1109/TED.2018.2876551.

[17] C. Mukherjee *et al.*, "Reliability-Aware Circuit Design Methodology for Beyond-5G Communication Systems," in *IEEE Transactions on Device and Materials Reliability*, vol. 17, no. 3, pp. 490–506, Sept. 2017, doi: 10.1109/TDMR.2017.2710303

[18] M. Schroter and A. Chakravorty, "Compact hierarchical modeling of bipolar transistors with HICUM", World Scientific, Singapore, ISBN 978-981-4273-21-3, 2010.

[19] M. Müller *et al.*, "Physics-Based Compact Modeling of the Transfer Current in III-V DHBTs with the Generalized Integral Charge Control Relation," *2023 IEEE BiCMOS and Compound Semiconductor Integrated Circuits and Technology Symposium (BCICTS)*, Monterey, CA, USA, 2023, pp. 12–15, doi: 10.1109/BCICTS54660.2023.10310908.

[20] B. Ardouin *et al.*, "Compact Model Validation Strategies Based on Dedicated and Benchmark Circuit Blocks for the mm-Wave Frequency Range," *2015 IEEE Compound Semiconductor Integrated Circuit Symposium (CSICS)*, New Orleans, LA, USA, 2015, pp. 1–4, doi: 10.1109/CSICS.2015.7314492.

Bertrand Ardouin received the M.S. and Ph.D. degrees in electrical engineering from the University of Bordeaux, Bordeaux, France, in 1998 and 2001, respectively. In 2002, he was the co-founder of XMOD Technologies, Bordeaux, a startup he has led as a CEO until 2019. After the acquisition of XMOD by SERMA Group, he was a Business Unit Manager with SERMA Technologies, Pessac, France, in charge of electrical expertise and compact modeling activities. He joined Nokia Bell Laboratories and III-V Lab, where he is currently responsible of the High-Speed Analogue Digital Interfaces Group, in Palaiseau, France. His areas of research interest cover CMOS, SiGe, and InP HBT devices, harsh-environment electronics, reliability, RF and mmWave characterization, modeling and PDK, process technology, and high-speed IC design.

Tom K. Johansen (Member, IEEE) received the M.S. and Ph.D. degrees in electrical engineering from the Technical University of Denmark, Kongens Lyngby, Denmark, in 1999 and 2003, respectively. In 1999, he joined the Electromagnetic Systems Group, DTU Elektro, Technical University of Denmark, where he is currently an associate professor. From September 2001 to March 2002, he was a visiting scholar with the Center for Wireless Communication, University of San Diego, San Diego, CA, USA. He has spent several external research stays with the Ferdinand-Braun-Institut (FBH), Berlin, Germany. His current research interests include the modeling of high-frequency solid-state devices, and microwave, mmWave, and submillimeter-wave integrated circuit design. Dr. Johansen is a member of the IEEE.

Antoine Chauvet (Student Member, IEEE) received the Engineering degree in electrical engineering from the ENSEA, Cergy, France, in 2022. In 2023, he joined Nokia Bell Laboratories and III-V Lab, Nozay, France, as a Ph.D. student. His work focuses on the design and characterization of high-symbol-rate InP-DHBT integrated circuits for over 1 Tb/s/channel optical communications, and on trans-impedance amplifiers implemented into photoreceivers.

Romain Hersent (Member, IEEE) received the Engineering degree in electrical engineering from the ENSEA, Cergy, France, in 2016, and the Ph.D. degree in electrical engineering from L'Université de Cergy-Pontoise, Cergy, in 2020. He joined Nokia Bell Laboratories and III-V Lab, Nozay, France, in 2020, where he is currently a research engineer. His work focuses on the design and characterization of high-symbol-rate large-output-swing InP-DHBT integrated circuits for over 1 Tb/s/channel optical communications, and more specifically on analog multiplexers and linear drivers. Dr. Hersent is a member of the IEEE.

Virginie Nodjiadjim (Member, IEEE) received the Ph.D. degree in electronic engineering from the University of Lille I, Lille, France, in 2009. In 2009, she joined III-V Lab as a research engineer, where she has studied compound semiconductor heterojunction bipolar transistors, and is currently in charge of the InP-DHBT development through the optimization of the structure, the design of the device, characterization, and modeling. Dr. Nodjiadjim is a member of the IEEE.

Agnieszka Konczykowska (Life Fellow, IEEE) received the Ph.D. degree in electrical engineering from the Warsaw University of Technology, Warsaw, Poland, in 1977. She started her scientific activity with work on computer-aided design (CAD) tools and design methodologies. She continued expanding to device modeling and characterization and to circuit design. From 2005 to 2018, she was in charge of microelectronic design activity at III-V Lab. She is currently with ADesign in the domain of components and integrated circuits for telecommunication systems. From 1995 to 1999, she was the President of European Circuit Society. Dr. Konczykowska was elevated to an IEEE Fellow for "contributions to development of very high-speed circuits" in 2018. Since 2019, she has been an Editorial Board Member of Proceedings of the IEEE.

Nil Davy (Member, IEEE) received the Engineering degree in electrical engineering from Phelma, Grenoble, France, in 2020, and the Ph.D. degree in electrical engineering from the University of Bordeaux, Bordeaux, in 2024. In 2023, he joined Nokia Bell Laboratories and III-V Lab, Palaiseau, France, where he is currently a research scientist. His current research interests include the fabrication, design, and characterization of InP double heterojunction bipolar transistors.

Muriel Riet was born in Choisy-le-Roi, France, in 1958. She received the Ph.D. degree in electronic engineering from the University of Paris XI, Bures-sur-Yvette, France, in 1985. She joined CNET, Research Center of France TELECOM, Rennes, France, in 1985, where she has been in charge of HBT technology for high bit-rate optical communications up to 40 Gb/s. She has been in charge of InP HBT technology for high-bite-rate optical commutations up to 40 Gb/s with the III-V Lab, Palaiseau, France, since 1998.

Colin Mismer received the M.Sc. (engineering) degree in chemical industry from the ENSICAEN, Caen, France, in 2011. He joined Nokia Bell Laboratories and III-V Lab, Palaiseau, France, in 2017, where he is currently a Microelectronics Process Engineer. His work focuses on the fabrication of InP-DHBT integrated circuit and the development of new technological building blocks.

Analysis of Quasi-Coaxial Via Implemented in IC Substrate using Multiple-Scattering Method

Hiroaki Takahashi, Pelin Suealp, and Erich Schlaffer

Abstract—**This paper presents an analytical study of a quasi-coaxial via implemented in an IC substrate using the multiple-scattering method. The plate-via impedance matrix is derived using addition theorems for cylindrical inward and outward waves around the vias. This matrix is then utilized to determine the loop inductance, taking into account the via-barrel radius. The derived equations facilitate the investigation of loop inductance, which serves as a figure of merit for the quasi-coaxial configuration. Additionally, the paper includes a statistical analysis that accounts for design parameter uncertainties from via manufacturing point of view.**

Index Terms—**Via modeling, quasi-coaxial via, multiple-scattering method, loop inductance**

I. INTRODUCTION

Wireless communication with high data rates is the indispensable infrastructure for cutting-edge technologies such as artificial intelligence, machine learning (ML), virtual reality, quantum communication, quantum ML, blockchain, THz-sensing, and edge computing [1]. To fulfill the increasing demand for fast and low-latency communication, millimeter-wave bands are available with the launch of the 5G New Radio based on advanced semiconductor technologies. Further enhancement in network infrastructure with the concept beyond 5G/6G seeks for the possible use of communication using frequencies above 100 GHz as D-band (110–170 GHz) shows great potential and is being investigated for emerging future wireless communication technologies [2].

Advanced packaging technologies play a crucial role in the next generation of wireless communications. In the early stages of the chip and antenna design phases, packaging systems have to be co-developed and tested as it is critical due to the challenges of system integration [3]. To increase the degree of integration density for electrical interconnections, embedding technologies have been proposed and demonstrated. For instance, the so-called center core embedding presented in [4] uses organic IC-substrates or printed-circuit boards (PCBs) to place active and passive components inside the structures. The embedding of multiple components can be realized through panel-level technologies with chip-first embedding or chip-last assembly. This contributes to the further small footprint, high mechanical stability, and reliable copper interconnections without soldering as shown in Figure 1.

Regarding electrical performance, ohmic losses derived from lossy dielectric material and conductor material with

Research & Development, AT & S Austria Technologie & Systemtechnik Leoben, Austria
{h.takahashi, E.Schlaffer}@ats.net

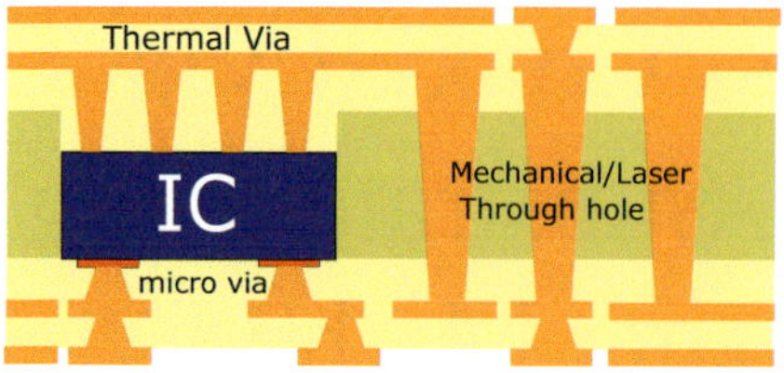

Fig. 1. Stack-up of a center core embedding with different via interconnections: microvia, thermal via, and mechanical/laser through hole.

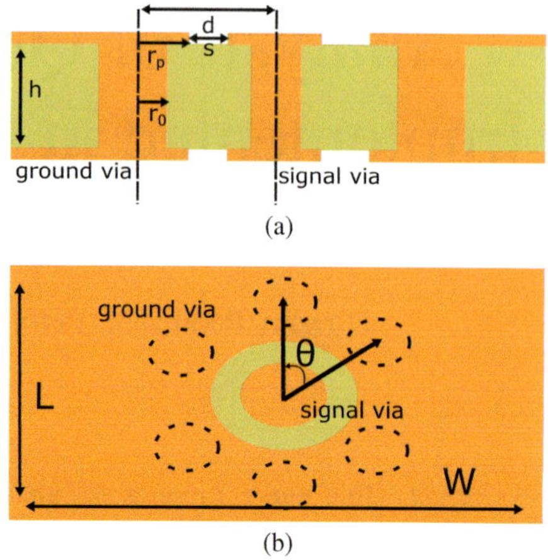

Fig. 2. Cross-section of quasi-coaxial via: (a) xz-plane; (b) xy-plane.

finite conductivity increase at higher frequencies. On the other hand, inductive and capacitive parasitic effects significantly impact signal integrity at higher frequencies. These parasitic effects are introduced at discontinuities, which are common in complex IC packages. Such discontinuities occur at curves, bends, and, most notably, vertical interconnections involving via structures. Vias enable vertical interconnections across IC substrates, from embedded chips to PCBs. However, without careful design, it can lead to significant degradation of electrical performance.

The electrical behavior of via structures has been investigated from an electromagnetic compatibility perspective through both theoretical and experimental approaches. A vast amount of hybrid field-circuit via modeling has been previously presented [5]–[11]. A via located within a parallel conductor plate can be seen as equivalent to a radial waveguide with transverse magnetic modes to propagation direction z, denoted as TM^z_{mn} [10]. Moreover, the currents flowing at the surface of a via barrel cause the no-cut-off parallel plate propagation wave and evanescent waves between the via and the plate contributing to the increase of the reactance [7]. The excitations can result in additional loss, reflection, and unwanted coupling to other signal lines. To improve signal integrity at the vertical interconnection, multiple ground vias surrounding the signal via can be utilized to suppress spurious emissions and undesired reflections [12], which is named quasi-coaxial via. The authors in [13] and [14] demonstrated the analysis of the loop inductance derived from the quasi-coaxial via based on an analytical inductance formula or physics-based via modeling approach. The loop inductance should be a figure of merit to clarify the frequency-dependent performance of quasi-coaxial via. In this study, we exploited the multiple-scattering method presented in [5] and [6] for the loop inductance analysis to take the via barrel radius into account, which is ignored in the previous studies. Furthermore, statistical analysis of the loop inductance based on the manufacturing capability of vias implemented in a typical IC substrate is also presented. The structure of the paper is as follows:

Section II briefly explains the theory behind the study, the extraction of loop inductance of via, and the calculation of via plate impedance matrix using the multiple-scattering method. Section III demonstrates the loop inductance analysis based on several cases of quasi-coaxial via as well as statistical analysis from the manufacturing point of view, which is followed by the conclusion in Section IV.

II. THEORY

In this section, the structure to be analyzed is introduced, and the procedure for the extraction of loop inductance and the derivation of plate via impedance matrix using the multiple-scattering approach are introduced. The structure to be analyzed is shown in Figure 2. Quasi-coaxial via consists of a signal via and surrounding ground vias. Design parameter is the following: the via barrel radius r_0, the via pad r_p, the distance to the edges s, total pitch d, and the height of the dielectric layer h. Due to the ease of theoretical derivation, the dielectric layer is assumed to be homogeneous over the domain of interest with identical relative dielectric constant, ε_r or Dk, and loss tangent, tanΔ or Df.

For signal integrity analysis of the quasi-coaxial via structure, the concept of the plate via impedance matrix denoted as $\mathbf{Z}_{\mathrm{PP}}$ is used, which is formulated in

$$\mathbf{V} = \mathbf{Z}_{pp}\mathbf{I}, \qquad (1)$$

where $\mathbf{V} = [V_1, V_2, \ldots, V_{\mathrm{P}}]^T$, and $\mathbf{I} = [I_1, I_2, \ldots, I_{\mathrm{P}}]^T$. P is the number of vias including signal via and ground vias. This P × P matrix relates the voltage of a via to the current of the via and the others. Ground vias are defined as the via shorted to the ground plate, i.e., $V_i = 0$. According to [13] and [14], the plate via impedance matrix $\mathbf{Z}_{\mathrm{PP}}$ can be converted into the loop inductance matrix denoted as $\mathbf{L}_{\mathrm{Loop}}$ depending on the number of signals and ground vias.

A. Extraction of Loop Inductance

Although the conversion method is detailed in [13] and [14], the mathematical procedure is briefly explained as follows: To determine the loop inductance matrix $\mathbf{L}_{\mathrm{Loop}}$, we exploit the imaginary part of the via-plate impedance matrix, $\mathbf{Z}_{\mathrm{PP}} = \mathbf{R} + j\omega\mathbf{L}_{\mathrm{PP}}$ as

$$\mathbf{V} = j\omega\mathbf{L}_{\mathrm{pp}}\mathbf{I}. \qquad (2)$$

To derive the loop inductance, the plate via inductance matrix is decomposed into blocks, namely signal inductance matrix $\mathbf{L}_{\mathrm{SS}}$, ground inductance matrix $\mathbf{L}_{\mathrm{GG}}$, and signal-ground matrix $\mathbf{L}_{\mathrm{SG}}$ and $\mathbf{L}_{\mathrm{GS}}$ as shown below.

$$\begin{bmatrix} \mathbf{V}_S \\ \mathbf{V}_G \end{bmatrix} = j\omega \begin{bmatrix} \mathbf{L}_{\mathrm{SS}} & \mathbf{L}_{\mathrm{SG}} \\ \mathbf{L}_{\mathrm{GS}} & \mathbf{L}_{\mathrm{GG}} \end{bmatrix} \begin{bmatrix} \mathbf{I}_{\mathrm{S}} \\ \mathbf{I}_{\mathrm{G}} \end{bmatrix}, \qquad (1)$$

where $\mathbf{V}_S = [V_1, V_2, \ldots, V_S]^T$, and $\mathbf{V}_G = [V_{S+1}, V_{S+2}, \ldots, V_G]^T$, as well as $\mathbf{I}_S$ and $\mathbf{I}_G$. Taking the inverse of the inductance matrix $\mathbf{L}^{-1} = \mathbf{B}_{\mathrm{PP}}$ and considering the relation between signal and ground vias, i.e., $\Sigma_{i=1}^{S} I_i = -\Sigma_{j=1}^{G} I_j$, the dimension of the matrix can be reduced from P × P to (S + 1) × (S + 1) as follows:

$$\begin{bmatrix} \mathbf{B}_{\mathrm{SS}} & \mathbf{B}_{\mathrm{SG}}\mathbf{e}_{\mathrm{G}} \\ \mathbf{e}_{\mathrm{G}}^T\mathbf{B}_{\mathrm{GS}} & \mathbf{e}_{\mathrm{G}}^T\mathbf{B}_{\mathrm{GG}}\mathbf{e}_{\mathrm{G}} \end{bmatrix} \begin{bmatrix} \mathbf{V}_S \\ V_G \end{bmatrix} = j\omega \begin{bmatrix} \mathbf{I}_S \\ I_G \end{bmatrix} \qquad (3)$$

where $\mathbf{e}_G$ is the $G \times 1$ column vector whose entries are all ones. Taking inverse of eqn (3), we can obtain the equivalent inductance matrix $\mathbf{L}_{\mathrm{eq}}$, from which the loop inductance matrix $\mathbf{L}_{\mathrm{Loop}}$ is calculated,

$$\begin{bmatrix} \mathbf{V}_S \\ V_G \end{bmatrix} = j\omega \begin{bmatrix} \mathbf{L}'_{SS} & \mathbf{L}'_{SG} \\ \mathbf{L}'_{GS} & \mathbf{L}'_{GG} \end{bmatrix} \begin{bmatrix} \mathbf{I}_S \\ I_G \end{bmatrix} \qquad (3)$$

$$\mathbf{L}_{\mathrm{Loop}} = L_{SS} - L_{SG}\mathbf{e}_G^T - \mathbf{e}_s L_{GS} + L_{GG}\mathbf{e}_G\mathbf{e}_S^T \qquad (4)$$

where $\mathbf{e}_S$ is the $S \times 1$ column vector whose entries are all ones.

B. Multiple-Scattering Method

As formulated, the loop inductance of quasi-coaxial vias is extracted from the via-plate impedance matrix $\mathbf{Z}_{PP}$. This indicates that the accuracy of the analysis depends on that of the calculation for $\mathbf{Z}_{PP}$. There have been several methods proposed to determine Z_{PP}, such as physics-based via modeling [7], intrinsic via modeling [9], and multiple-scattering method [5], [6]. The multiple-scattering method enforces the boundary conditions at the surfaces of via barrels and takes the electromagnetic field interactions between vias in the parallel plate propagation mode and evanescent modes into account. A quasi-coaxial via is supposed to have electrical characteristics dependent on their geometry and relative location between signal via and ground vias. Thus, the multiple-scattering method is employed for rigorous analysis. In the original multiple-scattering method, one needs to solve the Foldy–Lax equation including complicated Green's functions. The theory is reformulated by Zhang et al. in [6] to a simplified form and the mathematical procedure for Z_{PP} is briefly explained as follows:

The method considers the wave distributed around an ith via and the parallel plate decomposed into an infinite summation of modes as below

$$\mathbf{E}_Z^{(i)} = \sum_{m=-\infty}^{\infty}\sum_{n=0}^{\infty}\left\{\begin{matrix} b_{mn}^{(i)}(\rho^i,\phi^i,z) \\ +a_{mn}^{(i)}J_{mn}^{(i)}(\rho^i,\phi^i,z)\end{matrix}\right\}, \quad (4)$$

where ρ^i, ϕ^i, z are cylindrical local coordinates of an ith via, and $b_{mn}^{(i)}$ and $a_{mn}^{(i)}$ are the mode expansion coefficients of outward and inward cylindrical harmonics, respectively. It is worth mentioning that the dielectric layer thickness should be enough small that TE_{mn}^z modes can be ignored. The cylindrical harmonics $H_{mn}^{(i)}$ and $J_{mn}^{(i)}$ are respectively described as

$$J_{mn}^{(i)}(\rho^i,\phi^i,z) = J_m(k_n\rho^i)e^{jm\phi^i}\cos\left(\frac{n\pi}{h}z\right) \quad (5)$$

$$H_{mn}^{(i)}(\rho^i,\phi^i,z) = H_m^{(2)}(k_n\rho^i)e^{jm\phi^i}\cos\left(\frac{n\pi}{h}z\right), \quad (6)$$

where k_n is the effective wavenumber, i.e., $k_n = \sqrt{\varepsilon_r k_0^2 - \left(\frac{n\pi}{h}\right)^2}$, and $J(\cdot)$ and $H_m^{(2)}(\cdot)$ are the mth-order Bessel function of the first kind and mth-order Hankel function of the second kind, respectively. As detailed in [9], the accuracy of the analysis depends on the truncated number M for the azimuthal component and N for the axial component. When n equals to zero, the no-cutoff parallel plate propagating wave can be taken as the wave scattered at each via, whereas evanescent waves are considered with $n \geq 1$. In this study, m is always zero because the distance between vias is so large due to the constraint by the laser via the manufacturing process that higher-order azimuthal components can be ignored. Considering the mathematical assumptions above, one can reformulate the equation of E_z and based on Maxwell's equations, $H_\phi(\rho^i, \phi^i, z)$ can be also derived as

$$H_\phi^{(i)} = \sum_{n=0}^{N}\frac{j\omega\epsilon}{k_n}\begin{bmatrix} b_{0n}^{(i)}H_1^{(2)}(k_n\rho^i) \\ +a_{0n}^{(i)}J_1(k_n\rho^i)\end{bmatrix}\cos\left(\frac{n\pi}{h}z\right). \quad (7)$$

Therefore, the total magnetic field around quasi-coaxial via is described as

$$H_\phi(r,\phi,z) = \sum_{i=1}^{P}H_\phi^{(i)}(\rho^i,\phi^i,z). \quad (8)$$

The inward mode expansion coefficients of each via, $a_{0n}^{(i)}$ are included in the mode expansion coefficient vectors $\mathbf{a}_{0n} = [a_{0n}^1, a_{0n}^2, \ldots, a_{0n}^P]$ as well as $b_{0n}^{(i)}$. According to [9], the mode expansion coefficient vectors can be determined by addition theorems of cylindrical waves, including both the zero-order propagating and the higher-order evanescent modes as

$$\mathbf{a}_{0n} = \left(\mathbf{I} - \mathbf{S}_{PP}^R\Gamma\right)^{-1}\left(\mathbf{a}_{en} + \mathbf{S}_{PP}^R\mathbf{b}_{en}\right) \quad (9)$$

$$\mathbf{b}_{0n} = \Gamma \mathbf{a}_{0n}, \quad (10)$$

where $\mathbf{a}_{en}$ and $\mathbf{b}_{en}$ represent the expansion coefficients of the initial inward and outward cylindrical waves derived from the excited TEM (Transverse Electromagnetic mode) at the excitation ports as

$$b_{en}\left(V_0, z'\right) = \frac{j\pi V_0}{h\ln r_p^i / r_0^i} \frac{\cos\left(\frac{n\pi z'}{h}\right)}{1+\delta_{n0}} \left[J_0\left(k_n r_p^i\right) - J_0^{(2)}\left(k_n r_0^i\right)\right] \quad (11)$$

$$a_{en}\left(V_0, z'\right) = \frac{j\pi V_0}{h\ln r_p^i / r_0^i} \frac{\cos\left(\frac{n\pi z'}{h}\right)}{1+\delta_{n0}} \left[H_0\left(k_n r_p^i\right) - H_0^{(2)}\left(k_n r_0^i\right)\right], \quad (12)$$

and $\mathbf{H}_0 = \mathrm{diag}\left\{H_0^{(2)}(kr_i)h\right\}$, $\mathbf{J}_0 = \mathrm{diag}$ $\{J_0(kr_i)h\}$, $\mathbf{H}_1 = \mathrm{diag}\left\{\frac{2\pi k r_i}{j\omega\mu} H_1^{(2)}(kr_i)\right\}$ $\mathbf{J}_1 = \mathrm{diag}\left\{\frac{2\pi k r_i}{j\omega\mu} J_1(kr_i)\right\}$, $\Gamma_0 = \mathrm{diag}$ $\left[-\frac{J_0(k_n r_0^i)}{H_0^{(2)}(k_n r_0^i)}\right]$ and the element of $\mathbf{S}_{\mathrm{PP}}^R$ can be calculated as

$$S_{\mathrm{PP}}\left(i,j\right) = \left(1-\delta_{ij}\right) H_0^{(2)}\left(k_n r_{ij}\right), \quad (13)$$

where r_{ij} means the distance between vias. Using the matrices above, the via plate impedance matrix derived from Multiple scattering method can be also formulated as

$$\mathbf{Z}_{\mathrm{PP}}^R = \left(\mathbf{H}_0 + \mathbf{J}_0 \mathbf{S}_{\mathrm{PP}}^R\right)\left(\mathbf{H}_1 - \mathbf{J}_1 \mathbf{S}_{\mathrm{PP}}^R\right)^{-1}. \quad (14)$$

III. ANALYSIS

In this section, using the derived analytical equations in Sections II-A and II-B, the loop inductance of quasi-coaxial via dependent on the via geometry, the distance between vias, and the number of ground vias is investigated after the verification of analysis by comparison to the numerical simulation.

A. Verification of the Analytical Magnetic Fields

For verification of the analysis presented in the last section, the total magnetic field distribution $|H_\phi(r, \phi, z)|$ at 100 GHz around a quasi-coaxial via implemented in a homogeneous dielectric layer is calculated in Python/Julia environment and compared to the result of full-wave simulation. The dielectric material is assumed to be one of the typical materials for high-frequency IC-substrate, Panasonic LEXCMGX R-G545L, with Dk/Df of 3.6/0.002. Table 1 describes the parameters used for the analysis. Figure 3 shows the comparison in the normalized magnitude of magnetic fields between the equation and numerical simulation by CST Microwave Studio. Plot settings such as xy-range, scale, and contour levels are identical. The comparison shows a good agreement between the analysis and numerical simulation especially the H-fields with interaction between the signal via and ground vias by taking boundary conditions at via barrels rigorously into account.

B. Extraction of Loop Inductance of Quasi-Coaxial Via

As mentioned earlier, the loop inductance of signal via and ground vias is a

TABLE 1
MATERIAL AND GEOMETRICAL PARAMETER OF QUASI-COAXIAL VIA

h, µm	t, µm	W, mm	L, mm	Dk	Df
100.0	12.0	2.0	2.0	3.6	0.02

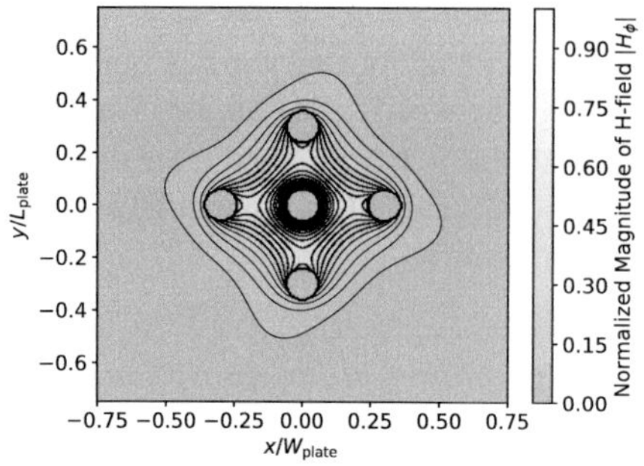

(a)

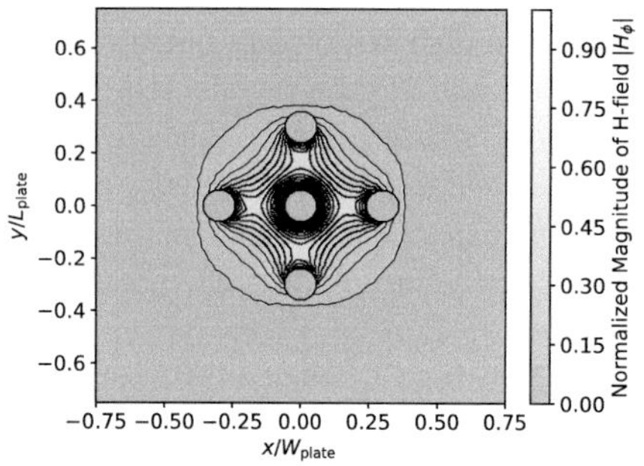

(b)

Fig. 3. Contour plot of the magnitude of magnetic fields around a signal via and four ground vias at 100 GHz (a). Analytical calculation by eqn (7) and (8) with N = 3 (b) full-wave simulation (CST).

function of the via alignment. This is due to the variation of H-fields to the via locations. Previous studies in [9] and [14] showed the fact that the loop inductance can be decreased by making the distance of vias small so that the magnetic field can be confined within the domain of quasi-coaxial structure. For example, suppose a signal via as via1 located at the center of the same plate used in Section III-A surrounded by two ground vias, via2 and via3, with identical dimensions. The location of via2 is fixed above via1 on the same y-axis at a distance of 250 μm. The loop inductance L_{Loop3}, in this case, depends on the location of via3. Figure 4 shows the contour plot of the loop inductance as a function of the location of via3, which is normalized to the case only

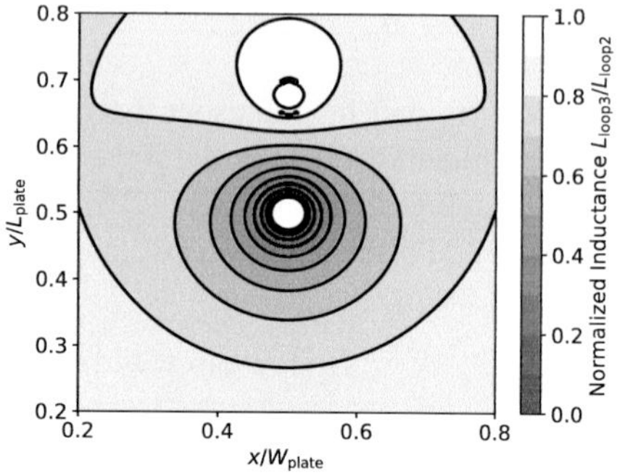

Fig. 4. Contour plot of normalized loop inductance of a quasi-coaxial via with three ground vias dependent on the location of the third ground via.

with via1 and via2 denoted as L_{Loop2}. The results suggest that the loop inductance varies with the position of via3 and can be reduced by positioning via3 closer to via1 and farther from via2. Similarly, Figure 5 illustrates the variation in loop inductance: (a) when the angle θ between two ground vias is varied at different distances from the signal via, and (b) when the distance from the signal via changes with different radii of the ground vias.

They implied that the loop inductance may be minimized when the distance between a signal via and ground vias is minimized and the distance between ground vias is maximized as well as the diameter of the ground vias.

C. Statistical Analysis of Quasi-Coaxial Via

The analysis of quasi-coaxial via so far clarified how the design parameters affect the loop inductance of the structures. Moreover, from a practical manufacturing point of view, there are some considerations to improve the manufacturing reliability that constrains the design parameters as listed below:

1. The area of the via pad with radius r_p should be sufficiently large due to

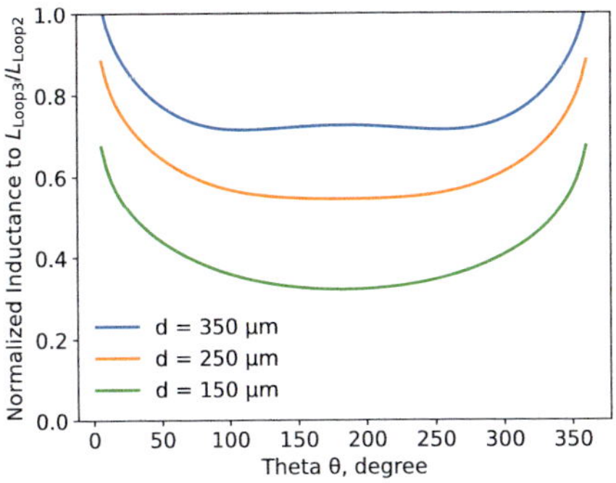

(a)

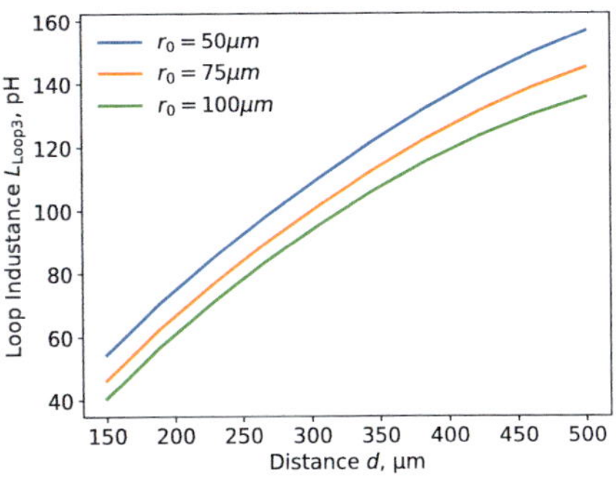

(b)

Fig. 5. The calculated loop inductance with a single via and two ground vias with different design parameters: (a) the angle θ between two ground vias as shown in Figure 2(b) is varied between [0, 2π] with different distances to the signal via; (b) the distance to the signal via is varied with different radius of the ground vias.

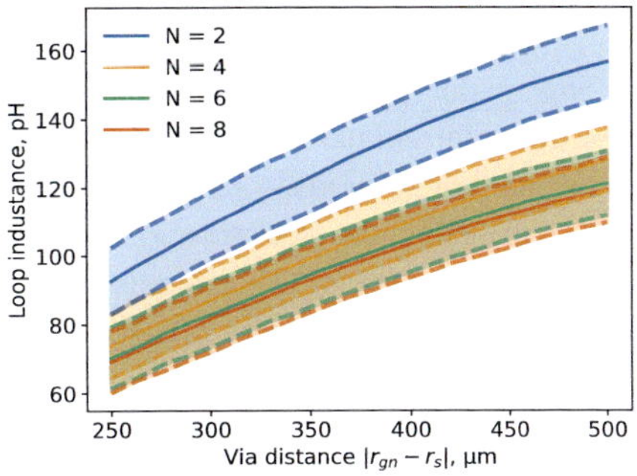

Fig. 6. The mean value of the loop inductance to the distance between a signal via and ground vias with different numbers of ground vias and their 95%- confidence interval of $2\sigma_L$ calculated from 1000 random Gaussian sampling based on input parameter variation $\sigma_r0 = 5$ μm and $\sigma_d = 5$ um

the misalignment during the drilling and lamination process.

2. Misalignment of via hole depends on the drilling process mainly by laser drilling or mechanical drilling.
3. Laser via hole diameter needs to be enlarged as the dielectric layer thickness increases.
4. Despite the better fabrication accuracy of laser drilling compared to mechanical drilling, it still suffers from misalignment and fabrication uncertainty of via radius due to the factors such as mechanical property of a substrate material, plating technology, lamination misalignment, laser focus adjustment, and pulse stability.

Therefore, statistical analysis that considers parameter uncertainties is essential for mass production. As mentioned earlier, the practical values of uncertainty parameters are highly dependent on the manufacturing process, drilling/plating technology, and dielectric material. In this study, the parameters with uncertainty include the via radius r_0, assumed to have a typical uncertainty of ±10 μm ($2\sigma_{r0}$) as well as the via distance d with ±10 μm ($2\sigma_d$). Figure 6 illustrates the loop inductance relative to the distance between a signal via and ground vias, incorporating various numbers of ground vias and including a 95% confidence interval calculated from 1000 Gaussian samples. As established in the previous section, the alignment of ground vias is symmetrical regardless of the number of ground vias present. The standard deviations of the loop inductance, σ_L, are 5.04 pH, 4.76 pH, 4.68 pH, and 4.69 pH when the number of ground vias is 2, 4, 6, and 8, respectively.

IV. CONCLUSION

This paper presented the analysis of quasi-coaxial via using the multiple-scattering method. The use of full-wave theory considering the inward and outward waves distributed via barrels by addition theorem enabled the rigorous calculation of plate-via impedance matrix. Afterwards, the derived equations were used for the study on the loop inductance of quasi-coaxial via dependent on the design parameters such as via radius, the distance between vias, and the their locations. Practical consideration in via manufacturing methods has been pointed out and, accordingly, statistical analysis was performed to figure out how the fabrication uncertainties can affect the loop inductance.

ACKNOWLEDGMENT

The financial support by Sustainable Technologies Enabling Future Telecom Applications (SHIFT) with project number 101096256-SHIFT-HORIZON-KDT-JU-2021-1-IA, delegated by the European Commission, is gratefully acknowledged.

REFERENCES

[1] M. W. Akhtar, S. A. Hassan, R. Ghaffar, H. Jung, S. Garg, and M. Shamim Hossain, "The shift to 6G communications: vision and requirements," *Human-centric Computing and Information Sciences*, Dec. 2020.

[2] X. Jia, X. Li, K.-s. Moon, J. W. Kim, K.-Q. Huang, M. B. Jordan, and M. Swaminathan, "Antenna-integrated, die-embedded glass package for 6g wireless applications," in *2022 IEEE 72nd Electronic Components and Technology Conference (ECTC)*, 2022, pp. 377–383.

[3] T. Maiwald, T. Li, G.-R. Hotopan, K. Kolb, K. Disch, J. Potschka, A. Haag, M. Dietz, B. Debaillie, T. Zwick, K. Aufinger, D. Ferling, R. Weigel, and A. Visweswaran, "A review of integrated systems and components for 6G wireless communication in the D-Band," *Proc. IEEE*, vol. 111, no. 3, pp. 220–256, Mar. 2023.

[4] M. Leitgeb and C. Vockenberger, *Die Integration Technologies on Advanced Substrates Including Embedding and Cavities*, 2022, pp. 193– 215.

[5] L. Tsang, H. Chen, C.-C. Huang, and V. Jandhyala, "Modeling of multiple scattering among vias in planar waveguides using foldy–lax equations," *Microwave and Optical Technology Letters*, vol. 31, no. 3, pp. 201–208, 2001. [Online]. Available: https: //onlinelibrary.wiley.com/doi/abs/10.1002/mop.1398

[6] Y. Zhang, J. Fan, A. R. Chada, and J. L. Drewniak, "A concise multiple scattering method for via array analysis in a circular plate pair," in *2008 Electrical Design of Advanced Packaging and Systems Symposium*. IEEE, Dec. 2008, pp. 143–146.

[7] Y. Zhang, J. Fan, G. Selli, M. Cocchini, and F. de Paulis, "Analytical evaluation of Via-Plate capacitance for multilayer printed circuit boards and packages," *IEEE Trans. Microw. Theory Tech.*, vol. 56, no. 9, pp. 2118–2128, Sep. 2008.

[8] R. Rimolo-Donadio, X. Gu, Y. H. Kwark, M. B. Ritter, B. Archambeault, F. de Paulis, Y. Zhang, J. Fan, H.-D. Bruns, and C. Schuster, "Physics- Based via and trace models for efficient link simulation on multilayer structures up to 40 GHz," *IEEE Trans. Microw. Theory Tech.*, vol. 57, no. 8, pp. 2072–2083, Aug. 2009.

[9] Y.-J. Zhang, G. Feng, and J. Fan, "A novel impedance definition of a parallel plate pair for an intrinsic via circuit

model," *IEEE Trans. Microw. Theory Tech.*, vol. 58, no. 12, pp. 3780–3789, 2010.

[10] S. Müller, X. Duan, M. Kotzev, Y.-J. Zhang, J. Fan, X. Gu, Y. H. Kwark, R. Rimolo-Donadio, H.-D. Brüns, and C. Schuster, "Accuracy of Physics-Based via models for simulation of dense via arrays," *IEEE Trans. Electromagn. Compat.*, vol. 54, no. 5, pp. 1125–1136, Oct. 2012.

[11] J. Fan, A. Hardock, R. Rimolo-Donadio, S. Müller, Y. H. Kwark, and C. Schuster, "Signal integrity: Efficient, physics-based via modeling: Return path, impedance, and stub effect control," *IEEE Electromagnetic Compatibility Magazine*, vol. 3, no. 1, pp. 76–84, 2014.

[12] D. Schwarz, N. Riese, A. Du¨rr, and C. Waldschmidt, "A broadband multilayer vertical transition at 79 GHz employing FR4 as core material," in *2021 51st European Microwave Conference (EuMC)*, Apr. 2022, pp. 700–703.

[13] J. Kim, L. Ren, and J. Fan, "Physics-Based inductance extraction for via arrays in parallel planes for power distribution network design," *IEEE Trans. Microw. Theory Tech.*, vol. 58, no. 9, pp. 2434–2447, Sep. 2010.

[14] P.-Y. Weng, C.-L. Liao, B. Mutnury, and T.-L. Wu, "Optimization of ground-via patterns for via transitions by minimizing loop inductance," *IEEE Transactions on Signal and Power Integrity*, vol. 2, pp. 43–52, 2023.

Hiroaki Takahashi received the B.E. (2016) and M.E. (2019) degrees from the University of Electro- Communications (UEC) Tokyo, Japan, and the Ph.D. (2023) in electrical engineering from Graz University of Technology, Graz, Austria. He joined R&D Department of AT&S Austria Technologie & Systemtechnik AG in 2022 and is currently working on physics-based modeling, characterization, and statistical analysis of microwave interconnections implemented in multilayered PCBs and IC substrates. He was the recipient of the Asia-Pacific Microwave Conference 2022 Student Prize. He is a member of IEEE, the Institute of Electronics, Information and Communication Engineering (IEICE), and the Japan Institute of Electronics Packaging (JIEP).

Pelin Suealp graduated with a bacehlor's degree in chemical and process engineering and master's degree in materials sciences from the Technical University of Vienna, Austria. She joined AT&S Austria Technologie & Systemtechnik AG in 2021 and was working until recently as R&D Senior Engineer.

Erich Schlaffer graduated from Technical Col- lege of Kapfenberg, Austria, in the field of electronics and has been with AT&S since 1998. He was a process engineer engaged in mechanical drilling, laser drilling, milling and electrical testing departments, as well as photo structuring. Since 2012, he has been a project manager in the R&D department, and responsible for the development of high frequency applications in the printed circuit board.

D-band Phased Array Antenna Module for 5G Backhaul

A. Fonte, S. Moscato, R. Moro, A. Pallotta, A. Mazzanti, A. Bilato, G. De Filippi, L. Piotto, F. Centurelli, P. Monsurrò, H. Sadeghi Chameh, P. Tommasino, A. Trifiletti, D. Lodi Rizzini, F. Tesolin, S. M. Dartizio, and S. Levantino

Abstract—**The 130–175 GHz frequency spectrum, identified as D-band, will be the key-enabler for next-generation backhauling point-to-point wireless links. Since capacity requirements for mobile networks will increase according to 5G/B5G deployment needs, a new paradigm for radio equipment needs to be developed. Integrated technologies based on silicon-germanium, exhibiting sub-THz cut-off frequency, are going to be adopted to design and implement the key building blocks for fully integrated transceivers. D-band radio frequency integrated circuits and antenna concepts are presented, tailoring phased array allowing electronic beamsteering and beamforming features.**

Index Terms—**5G, antenna array, backhaul, BiCMOS, D-band, low-noise amplifier, phased array, power amplifier, mmWave**

This work is supported by the Chips Joint Undertaking (Chips JU) under grant agreement number 101096256 (project SHIFT). The Chips JU receives support from the European Union's Horizon Europe research and innovation program and the National Authorities.

Alessandro Fonte, Riccardo Moro, and Stefano Moscato are with the R&D Department, SIAE MICROELETTRONICA, Cologno Monzese (MI), Italy (e-mail: alessandro.fonte@siaemic.com).

A. Pallotta is with STMicroelectronics s.r.l., v. Tolomeo, 1 Cornaredo (MI), Italy, A. Mazzanti, A. Bilato, G. De Filippi, and L. Piotto are with the Department of Industrial and Information Engineering, University of Pavia, Pavia, Italy, F. Centurelli, P. Monsurrò, H. Sadeghi Chameh, P. Tommasino, and A. Trifiletti are with Information Engineering, Electronics and Telecommunication Department, Sapienza University of Rome, Italy, D. Lodi Rizzini, F. Tesolin, S. M. Dartizio, and S. Levantino are with the Dipartimento di Ingegneria Elettronica, Informazione e Bioingegneria, Politecnico di Milano, Milano, Italy.

I. INTRODUCTION

THE current backhauling systems for mobile telecommunications allow for interconnection of distributed access stations to the core of the network. Wireless backhauling is still a competitive option over fiber optics, especially as multi 10 Gbps capabilities become available. Its share should indeed be maintained with the deployment of 5GNR and 6G, while the overall market will grow thanks to the increased demand for denser networks.

The developments of the technologies and systems carried out with this work aim to overcome the existing limitations and to facilitate the introduction of millimeter-wave (mmWave) telecommunications for the backhauling of telecommunications, replacing the current wireless links. This system is based on the antenna in module (AiM) solution, shown in Figure 1a and 1b, that provide alternative approaches with respect to the standard one (radio frequency integrated circuits (RFICs) and external

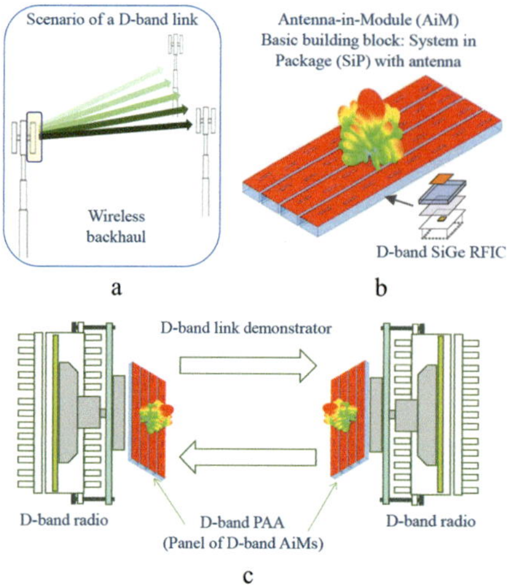

Fig. 1. Scenario of a D-band radio link (a), antenna in module (AiM) basic building blocks (antenna, package, and RFICs) (b), mockup of the D-band radio unit with integrated phased array antenna (PAA) (c).

antenna). Moreover, the AiM as part of a phased array antenna (PAA) increases the total output power through beamforming also allowing the possibility of steering functionality. The D-band AiM concept requires several challenges to be solved, and in particular, (i) at design level, high output power, and low-noise figure at D-bands can be reached, and a very high-integration of the RF front-end is required due to very limited area available for the dies in the PAA, (ii) at the package level, the AiM needs a launcher in the package (LiP) that is necessary for connection to the waveguide antenna.

The advantages of the AiM solutions are (i) the low-cost of the basic component, (ii) the plug and play on other wireless systems (off-the-shelf), and (iii) the reduced design time than any ad-hoc D-band solution.

This paper is organized as follows. In Section II, the D-band system architecture is explained. In Section III, the design of the D-band antenna of the AiM is described, and in Section IV the silicon-germanium (SiGe) RFICs, and in particular the key building blocks of the TX and RX D-band system are described. Finally, in Section V, the conclusions are drawn.

II. D-BAND SYSTEM ARCHITECTURE

The D-band panel of AiM, due to its small size, is expected to be replicated 4 or 8 times on the case of a practical equipment, such that 4 independent beams can be obtained. These can be exploited to offer multibeam equipment or a 2 × 2 MIMO (Multiple-input Multiple-Output) in dual polarization, thus allowing even more capacity and flexibility.

The AiM system will therefore enable the transition of backhauling/midhauling/fronthauling wireless links for 5G and future evolutions. The envisioned system architectures for communication systems involving the AiM are shown in Figure 1c. The relevant aspects of such architectures with the AiM components are: (i) the system designer should only be concerned with the core aspects of its application, such as data gathering, processing, user interaction, and can ideally effortlessly place the AiM components without dealing with mmWave RFICs design, (ii) the circuit designer should not be concerned by critical RF interconnections, which may often require a longstanding experience in the field (all connections needed are baseband ones), (iii) the device will not need an external antenna, as it is included into AiM components, and (iv) the application can rely on very-high-speed communications allowed by the usage of mmWaves and transparently handled by the AiM.

The AiM fully integrated module includes four TX/RX mmWave chains, designed in SiGe BiCMOS process,

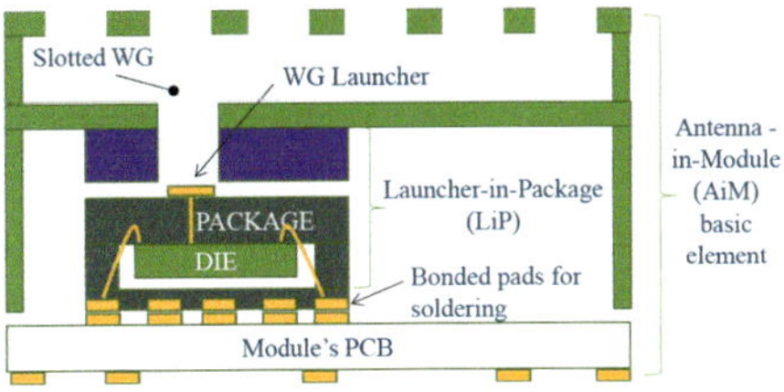

Fig. 2. First design of D-band AiM assembly.

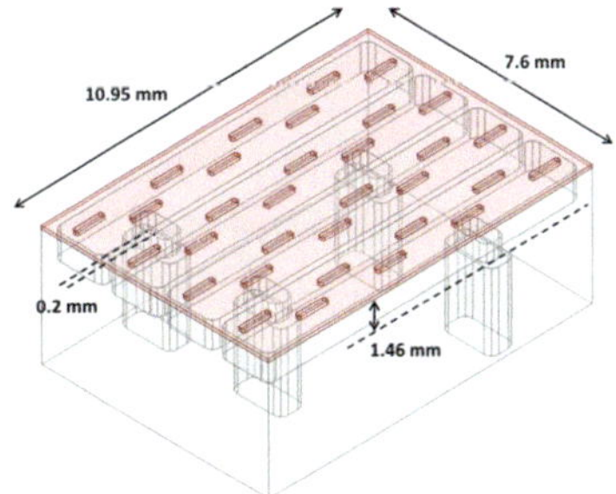

Fig. 3. Slotted waveguide D-band antenna structure.

together with biasing and control sections. For each chain, an LiP is required to feed the antenna located on the top of the structure. The aim of this architecture is to reach a very high level of integration that permits to reduce interconnection losses and to increase the system efficiency. Moreover, the module described exhibits all the characteristics of an active phase array since a proper tuning of the input signal can be performed. The phase and amplitude can be changed independently by the four mmWave chains to implement beamforming and beam-steering features. The first design of AiM assembly, shown in Figure 2, presents all the building blocks necessary to implement the D-band module.

Looking at the architecture, two main areas can be highlighted; the active components, i.e., the RFICs, and the antenna, whose characteristics and performance are described in the next sections.

III. ANTENNA-IN-MODULE DESIGN

The radiating structure and its integration in the AiM are important to limit the signal loss and to permit the scan of the beam, reducing the drawbacks tied to this approach (gain loss, grating lobes, etc.). In addition, a critical aspect to face in the antenna design is the size that cannot be excessive to keep the module replicable in more complex systems such as a MIMO configuration.

After a first analysis of several solutions presented in literature [1] and by converting the requirements described in an antenna typology, four in-line antenna arrays with a planar dimension limited to 8 × 16 mm^2 have been identified. The feeding points are positioned far away from each other to accommodate the RFICs. In terms of electrical performance, the antenna works in the D-band and has a minimum operating bandwidth of 10 GHz, which allows to address the maximum D-band channels bandwidth of 5 GHz and 20 dBi of minimum overall gain.

Among the different designs that meet those constraints, the most suitable one to guarantee good performance is based on the hollow metallic waveguide technology. An example that was developed and that can be used as an antenna reference for this system is the slotted waveguide antenna array shown in Figure 3. The structure consists of four blocks of eight radiating elements, which are linear slots etched in the waveguide top metal layer. The number of slots chosen allows reaching a gain high enough while keeping the tight dimensions but with the disadvantage of a narrow operating bandwidth. The position of the radiating elements was studied to achieve broadside linear polarized radiation at 145 GHz. Furthermore, the slots on the top layer together with

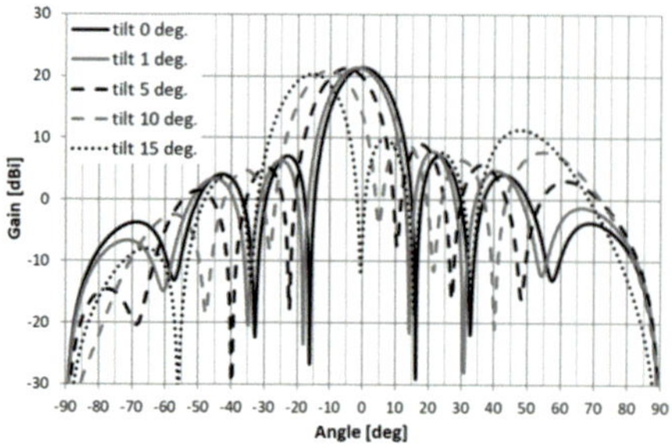

Fig. 4. Antenna radiation pattern for different steering angles.

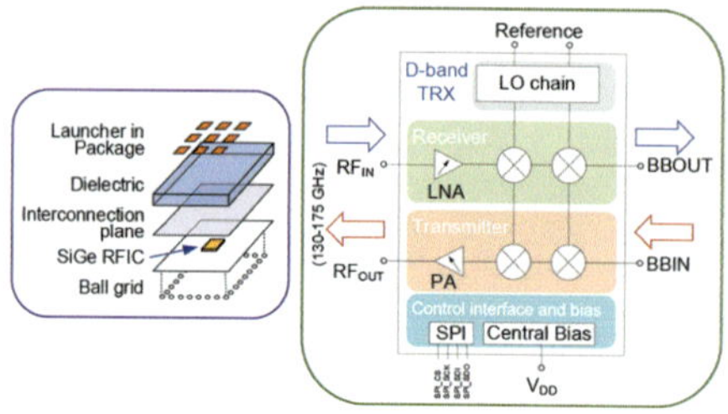

Fig. 5. Mockup of the package and the simplified building block scheme of the SiGe BiCMOS TRX RFIC.

the matching section in the feeding waveguide were optimized to guarantee an input return loss better than 15 dB.

The antenna array was simulated exciting the four input waveguides (standard WR6) in-phase. A 15 dB return loss is obtained over a bandwidth of 12 GHz (8% relative bandwidth). The input matching behavior is similar for all the ports showing that the impact of mutual couplings between the slots is negligible. At center frequency of 145 GHz, the antenna gain is 21.38 dBi. The simulation results confirm that the design presented is in line with the application requirements and it can be used as a reference for other structures that aim to improve its limitations. A wider bandwidth, for example, would be an added value also to have a margin for the manufacturing mechanical tolerances.

The presented structure has also beamsteering capability, which can be exploited to compensate for the misalignment and to simplify the deployment of the link. To verify the performance of the antenna for a displacement of a few degrees of the main lobe direction, a progressive phase on the consecutive ports was applied for different phase values. In Figure 4, the resulting E-plane radiation pattern is shown. Applying a phase shift of 10° between the four antennas, the main lobe is steered by 1.6° with a scan loss of 0.015 dB.

IV. D-BAND SIGE RFICS

This section presents the technology process selected for this project and some key active components of the D-band AiM: the power amplifier (PA), D-band mixers, the low-noise amplifier (LNA) and the low-phase-noise local oscillator (LO). A mockup of the package and the simplified building block scheme of the TRX RFIC are shown in Figure 5.

A. *Improved SiGe:BiCMOS HBT and Back-End Process*

The SiGe:BiCMOS technology has been proven as a viable and enabling technology for beyond 100 GHz mmWave front-end modules, by fully exploiting the flexibility and the integration capabilities envisioned by mixing digital and analog functionalities. SiGe Heterojunction Bipolar Transistor (HBT) architecture evolved significantly over the years with the objective of always improving (shrinking) the vertical profile (driving f_T) and reducing the parasitics (base resistance R_B and collector–base capacitance C_{BC} being critical for f_{MAX}). The exploitation of the radio spectrum in D-band (130–174.8 GHz), by relying on power-efficient RFIC chipset for an active phased antenna array system with beamsteering functionality, has been made feasible by an advanced

SiGe:BiCMOS process, based on a nine-metal stack back-end option and HBT structure having f_T and f_{MAX} > 400GHz.

The main concept behind the new process recipe is the use of a mono-crystalline base link between intrinsic SiGe base and the external base electrode to minimize the base resistance of the SiGe HBTs. Developments at STMicroelectronics started from the 55 nm BiCMOS platform currently in production, in which the high-speed SiGe HBT features 320 GHz f_T and 370 GHz f_{MAX}. The final objective is to offer a 400 GHz f_T and 500 GHz f_{MAX} HBT on the same CMOS node. Three major process changes are implemented to reach this objective [2]:

- The first one is the modification of the process thermal budget with the reduction of the spike annealing temperature and the addition of a millisecond annealing. It allows to increase the f_T of the SiGe HBT thanks to a reduction of dopants diffusion and a better activation of these dopants. Process conditions have also been defined to not degrade the performances of the CMOS transistors.
- The second change applies to the extrinsic collector module with the replacement of the buried layer by an implanted collector, which reduces both the cost and the cycle time of the technology. Super Shallow Trench Isolation (SSTI) is implemented to reduce the base–collector capacitance.
- Finally, a STMicroelectronics proprietary SiGe HBT architecture, called EXBIC (for Epitaxial eXtrinsic Base Isolated from the Collector), is being developed to address the f_{MAX} challenge. The key feature of this architecture is a boron in-situ doped epitaxial base link used to reduce the extrinsic base resistance.

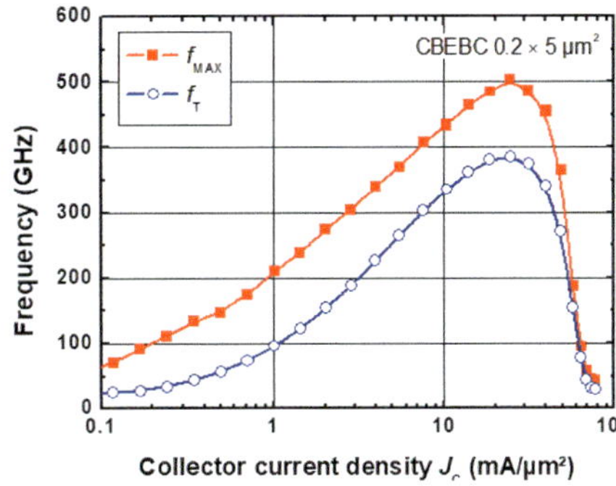

Fig. 6. Measured RF-performance (dots) versus SPICE model simulation (solid lines).

Measured RF-performances (f_T and f_{MAX}.vs. collector current density) are shown in Figure 6.

Besides the high-speed performances of the HBT, the nine-metal back-end of line (BEOL) has been adjusted to increase the Q-factor of inductors and reduce the attenuation constant of transmission lines (TLINEs). This is achieved by implementing two thick Cu layers for the vias (2.7 µm wide) and for the lines (2.3 µm wide) available in 300 mm wafers.

B. Heterodyne D-Band TX Front-End

The building block scheme of the transmission section of the D-band SiGe RFIC is shown in Figure 5. The two building blocks, i.e., the PA and the up-conversion mixer, were developed and validated in the STMicroelectronics 55 nm BiCMOS process and then redesigned in the new process with enhanced HBT transistor (i.e., the B55X process).

The D-band PA, whose circuit schematic is shown in Figure 7a, comprises HBTs in common base (CB) and a cascode input stage to rise the overall gain [3]. The CB configuration is selected, in most of the stages, giving its higher power and efficiency compared to the common emitter one. The HBTs area is scaled progressively along the chain, starting from the output stage that is sized to deliver a saturated power $P_{sat} > 17$ dBm. To improve the efficiency in back-off, the quiescent current is set low, such that the average current expands with the signal leveraging the current clamping mechanism, enabled by the CB stages with a DC-feed inductance from the emitter terminals to ground [4]. The matching networks are realized with TLINEs, allowing simpler and more reliable modeling of the return signal paths than spiral inductors, a critical issue in D-band. The TLINEs are shielded microstrips, with the signal in the topmost copper metal layer. The geometrical parameters (widths and spacings) are optimized to minimizing the TLINEs insertion loss.

Looking at the schematic of the up-conversion mixer in Figure 7b, [5], the transformer T_1 provides input matching for the IF cascode amplifier ($Q_{1,2}$–$Q_{3,4}$). The LO input is buffered with a single-ended cascode ($Q_{5,6}$) and made differential by T_3. The switching-quad ($Q_{7,8}$–$Q_{9,10}$) is fully balanced for LO suppression. It is found that a major limitation to the mixer output power stems from the low output resistance of the switching quad, reduced in D-band by the feedback of the HBTs Miller capacitors and the extrinsic base resistors. $Q_{11,12}$ are thus stacked to the switching quad to raise the output impedance and increase P_{out}. T_3 combines the output currents into a single-ended D-band signal. The HBTs in the switching quad, with emitters grounded by the secondary inductance of T_3, operate in CB for the IF signal. Thus, in the same way as in the PA, current clamping is exploited to expand the average current with the signal and maintain good efficiency in power back-off. The transformers T_1–T_4 are realized as coplanar windings in the topmost metal layer.

Photographs of PA and mixer in the BiCMOS-55 nm technology are shown in Figures 8 and 9, together with a measurement summary. The PA demonstrated 17.6 dBm P_{sat} with 2.2 V supply at 135 GHz. The small-signal gain is 24 dB with P_{out} at 1dB gain compression is $P_{1dB} = 16.8$ dBm, ~1dB below P_{sat}. The DC current rises from 30 mA at the quiescent point to 130 mA at P_{1dB}. The peak power-added efficiency (PAE) is 17.5% and still 8.5% at 6 dB power back-off. The mixer, with an IF signal at 40 GHz and –3 dBm LO at 100 GHz, achieved 15 dB conversion gain with RF and IF bandwidths of 38 and 15 GHz, respectively. The measured $P_{sat} = 6.3$ dBm and $P_{1dB} = 4.5$ dBm are remarkably high. The DC current rises from 34 mA at the quiescent point to 66 mA at P_{1dB}.

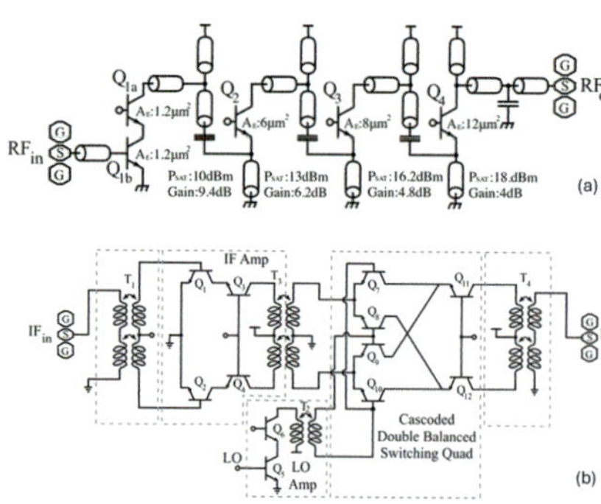

Fig. 7. Schematics of the D-band PA (a) and heterodyne up-conversion mixer (b).

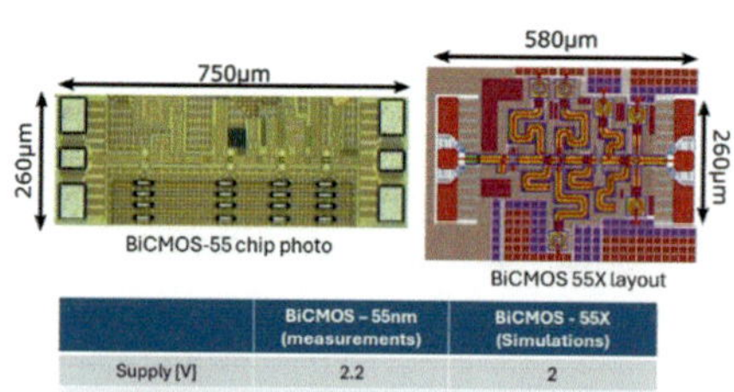

	BiCMOS – 55nm (measurements)	BiCMOS - 55X (Simulations)
Supply [V]	2.2	2
-3dB BW [GHz]	[125 - 159]	[110 – 175]
Gain [dB]	24	19
P_{SAT} [dBm]	17.6	[16.8 –19.8]
P_{1dB} [dBm]	16.8	[16.0 –19.5]
PAE [%]	17.5	[10.5-15.5]
PAE @ 6dB BO [%]	8.5	7

Fig. 8. D-band PA. Chip photograph in 55 nm BiCMOS, layout view in the new B55X process and performance summary and comparison.

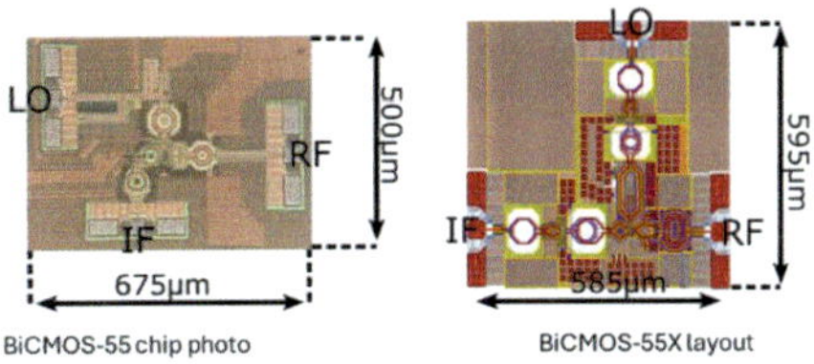

	BiCMOS – 55nm (measurements)	BiCMOS - 55X (simulations)
Supply [V]	2	2
-3dB RF-BW [GHz]	[112 - 150]	[126 – 175]
-3dB IF-BW [GHz]	[32-47]	[54-88]
LO power [dBm]	-3	-2
Conv-Gain [dB]	15	18.5
P_{SAT} / P_{1dB} [dBm]	6.3 / 4.5	6.8 / 5.5
η @ P_{1dB} [%]	2.5%	5.4%

Fig. 9. D-band up-conversion mixer. Chip photograph in 55 nm BiCMOS layout view in the new B55X process and performance summary and comparison.

The PA and mixer have been redesigned with performances adjusted to the target application, taking benefit of the advanced B55X process. The layout of the chips and the performance summary from post-layout simulations are reported in Figures 8 and 9. The layout of the PA was optimized with 25% reduced footprint. The gain is purposely reduced to 19 dB, considering the high P_{out} available from the mixer. Thanks to the higher HBTs f_T/f_{max}, the bandwidth is extended to 110–175 GHz, allowing to cover the full D-band with a single component. The expected P_{out} is comparable or slightly better than in the 55 nm BiCMOS design. The IF of the mixer is shifted to E-band and the operation bandwidth (both at RF and IF) is remarkably increased. The higher HBTs gain is exploited to raise the conversion gain with nearly the same P_{out} but remarkably higher expected efficiency.

C. Heterodyne D-Band RX Front-End

The receiving section of the D-band RFIC includes two D-to-E-band down-converters (D/C): a low-band (130–150 GHz) D/C and a high-band

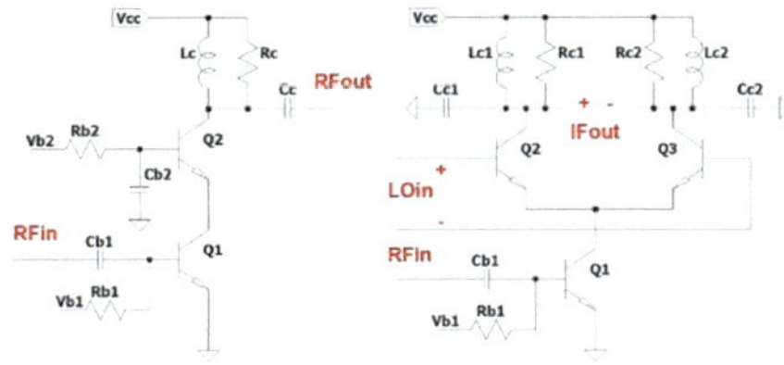

(a) (b)

Fig. 10. Simplified schematic of the LNA core (a) and of the D/C mixer core (b).

(150–165 GHz that can be extended to 175 GHz), which allow covering the overall D-band. The full D-band receiver will be obtained by cascading the D-band D/C stage to an E-band receiver already designed in 55 nm BiCMOS technology [6] and presently under design in the new B55X process.

The two RFICs implementing the two D/C blocks have been designed in B55X technology. They share the same architecture and circuital topology: they are composed of two building blocks, a two-stage D-band LNA and a single-balanced D/C mixer based on an active two-quadrant analog multiplier. The two above-mentioned building blocks have been also taped-out as stand-alone RFICs in the two sub-bands for testing purpose.

The core of the single-stage LNA has been designed exploiting a cascode topology, as shown in the simplified circuit schematic of Figure 10a: the load inductor is implemented with a minimum-width microstrip line, provided by the Process Design Kit (PDK). The cascode stage bias current has been chosen in order to guarantee a low-noise Figure (< 8 dB) over the whole bandwidth, but without sacrificing the linearity performance (output P_{-1dB} approaching 0 dBm), according to the state-of-the-art of D-band BiCMOS LNAs [7].

Two versions of the LNA have been designed, the second one with the goal to increase the gain up to at least 20 dB.

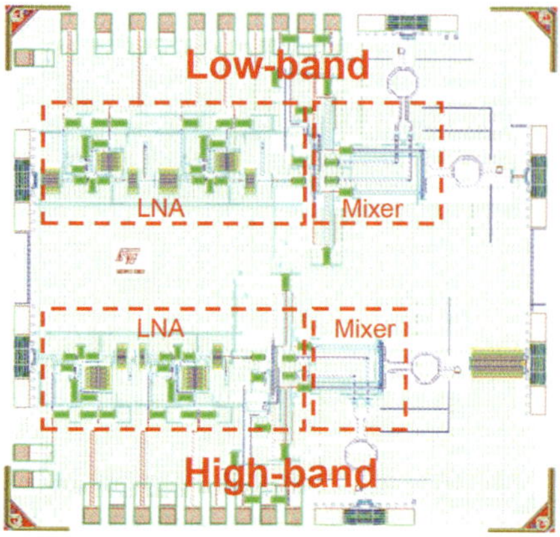

Fig. 11. Layout of the fabricated D-band to E-band D/Cs.

TABLE 1.
SIMULATED PERFORMANCE OF D-BAND BLOCKS.

	Gain [dB]	NF [dB]	OP_{1dB} [dBm]
Low-band LNA 1	17.5	6.6	−1.0
Low-band LNA 2	21.6	6.6	−3.4
High-band LNA 1	14.9	7.4	−2.0
High-band LNA 2	20.0	7.4	−3.5
Low-band mixer	−3.2	18.0	−4.2
High-band mixer	−5.5	23.0	−13.3
Low-band down-converter	14.0	7.6	−12.1
High-band down-converter	11.9	9.2	−13.0

The core of the down-conversion mixers, shown in Figure 10b, is a differential pair with a tuned *R-L-C* load. The tuned load is composed of a minimum-width microstrip line and a metal-oxide-metal (MOM) RF capacitor. For both the LNA and the mixer building block, the matching networks have been designed exploiting a single-stub topology: a microstrip line allows matching to 50 Ω of the real part. The remaining reactive part is cancelled by means of a grounded MOM capacitor or inductor (again implemented by using a minimum-width microstrip line). Finally, baluns have been used at the LO input and at the E-band IF output. Electro-magnetic post-layout simulations have been performed by Cadence EMX CAD tool to tune loads and matching networks, and also to evaluate effects of interconnections.

In Figure 11, the layout of the test chip containing the two D/C blocks is shown. The total area is 1.4 × 1.2 mm².

In Table I, post-layout simulation results are presented at the center of the bandwidth, i.e., 140 GHz for the low-band blocks and 157.5 GHz for high-band blocks. Simulated S-parameters and noise figure of the high-band LNA are shown in Figure 12.

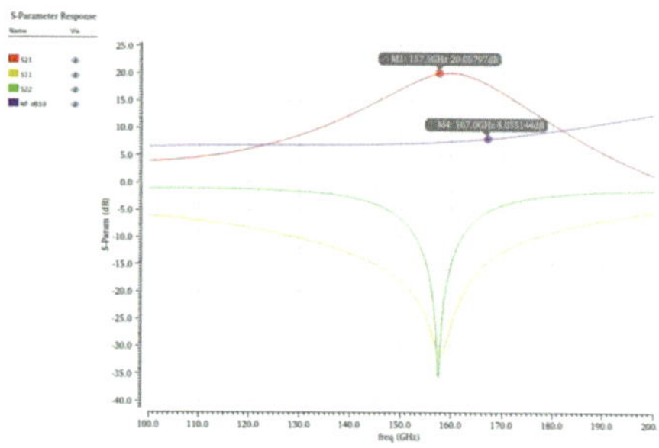

Fig. 12. Simulated S-parameters and noise figure of the high-band LNA.

D. *Low-Phase-Noise LO Generator*

The LO generator is intended to provide the LO signal to the D-band receiver. Considering the low-band D/C block with an input frequency in the 130–150 GHz range, an LO frequency spanning from 60 to 80 GHz allows to down-convert the signal to an IF frequency of 70 GHz. Frequency generation at mmWave is particularly challenging when low phase noise and wide tuning range (TR) are targeted. The straightforward way to generate a LO would be the direct frequency synthesis from a reference crystal oscillator by means of a phase-locked loop (PLL). However, the relatively low-quality factor of tunable resonators integrated on silicon substrates makes this solution

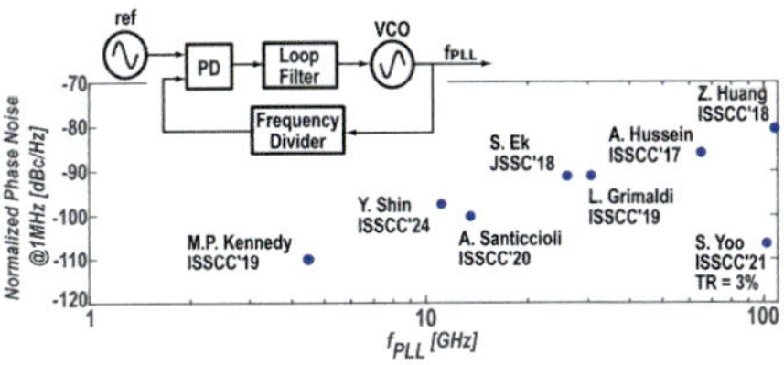

Fig. 13. Direct LO synthesis approach: block diagram of the LO generator and plot of the phase noise at 1 MHz offset from the carrier (normalized at 80 GHz) versus the VCO resonance frequency.

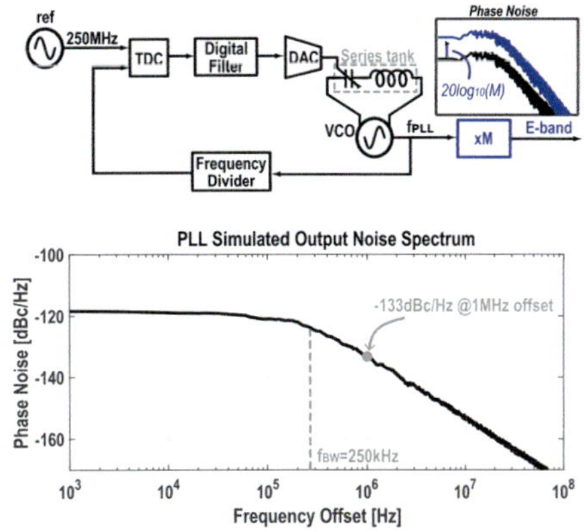

Fig. 14. Indirect LO synthesis approach through a DPLL (with series-resonance VCO) followed by a frequency multiplier, and simulated PLL spectrum.

impractical. This is evident looking at the state-of-the-art of PLL-based frequency synthesizers, whose phase noise presents a worsening trend as the voltage-controlled oscillator (VCO) resonance frequency f_{PLL} increases, as described by the plot of the PLL phase noise at 1 MHz offset from the carrier (normalized at a carrier frequency of 80 GHz) as a function of f_{PLL}, in Figure 13. The outlier is the work in [8], which achieves excellent phase-noise performances at an extremely narrow TR that would be impractical.

Therefore, an indirect frequency synthesis approach is typically preferred at mmWave, in which a PLL is followed by a frequency multiplier. The impact of the frequency multiplication stage on the output phase noise is simply an increase of the PLL phase noise level by . The adoption of frequency multipliers introduces other challenges related to their power consumption and harmonic rejection. In fact, the output of the multipliers contains the sub-harmonics of the output sinusoid. In this work, a good compromise is found by adopting a fractional-N PLL with an output frequency in the 10–13.33 GHz range and a 250 MHz reference signal (so that the PLL frequency division is limited to a factor comprised between 40 and 53.33), followed by a sextupler (performing a sixfold frequency multiplication) to get a local oscillation in the 60–80 GHz range. As shown in the block diagram in Figure 14, the PLL adopts a digital PLL (DPLL) architecture based on a time-to-digital converter (TDC) used as a phase detector to convert the time error between the reference signal and the one coming from the frequency divider into a digital word that is fed to a digital loop filter. The output of the filter is then converted to a voltage by means of a digital-to-analog converter (DAC) driving the VCO tuning voltage. DPLLs offer the big advantage of replacing the analog loop filter with a digital one, with great benefit in terms of area occupation (especially at narrow PLL bandwidths) and configurability [9].

Conventional VCOs based on parallel LC resonators have a voltage swing limited by transistor's breakdown. So, reaching ultra-low phase noise requires increasing power consumption by reducing the resonator impedance or, in other words, by decreasing the inductance L and increasing the capacitance C. However, this strategy has a limit that is reached when L is so small that becomes comparable with the parasitic inductance connecting the varactors. To overcome this limitation, the VCO adopts a configuration based on an integrated LC series resonator and HBT transistors, which has been recently introduced in [10]. This

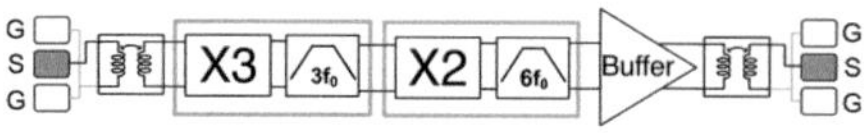

Fig. 15. Block diagram of the frequency sextupler.

solution allows to raise the power delivered to the resonator by increasing the current swing, with no need to scale down *L*. The VCO, designed in BiCMOS-55X process, has a 10–11.5 GHz (14%) TR, and reaches a simulated phase noise of −136 dBc/Hz at 1 MHz offset, with a DC power of 380 mW. A second VCO will be alternatively used to cover the remaining TR. Figure 2 shows the simulated output spectrum of the overall PLL, including the phase noise of reference buffer, TDC, DAC, and frequency divider. The PLL bandwidth f_{BW} of about 250 kHz is chosen to minimize the integrated rms phase noise, which is of about –63 dBc (equivalent to an absolute jitter of 16 fs rms), while the phase noise at 1 MHz offset at the PLL output is of – 133 dBc/Hz.

The sextupler is implemented in B55X as the cascade of a tripler and a doubler, as shown in Figure 15. The two blocks adopt the circuit topology disclosed in [11]. The sextupler achieves, from post-layout simulations, a harmonic rejection better than 29 dB across the output frequency range from 60 to 78 GHz. It has a DC power consumption of 67.3 mW and an area occupation of 0.32 mm^2. The simulated phase noise at the output of the cascade of the PLL and the sextupler is about −117 dBc/Hz at 1 MHz offset.

V. CONCLUSION

The advancement on the development of the basic element of the D-band (130–174.8 GHz) PAA system, called AiM, has been presented. The system architecture, the mockup of the module and the package, the D-band antenna concept, and the RFICs developed by using the cutting-edge STMicroelectronics 55 nm and 55X SiGe BiCMOS technologies have been described and the first results have been shown.

The D-band antenna has been designed to achieve broadside linear polarized radiation at 145 GHz, 15 dB return loss over a bandwidth of 12 GHz, and a gain greater than 20 dBi. Regarding the transmitter section, simulations (in SiGe B55X process) and measurement (in 55 nm SiGe BiCMOS process) results have shown that the D-band PA can achieve power gain > 19 dB, P_{SAT} > 17 dBm and PAE between 10.5% and 17.5%. The D-band up-conversion mixer can achieve conversion gain > 15 dB, P_{SAT} > 4.5 dB, and an operating bandwidth covering the whole D-band. Regarding the receiver section, simulation results (in SiGe B55X process) have shown that the D-band LNA can achieve power gain > 15 dB, noise figure better than 7.4 dB and OP1dB > −4 dBm. The D-band down-conversion mixer can achieve conversion gain > 12 dB, noise figure < 9.2 dB and OP1dB > −13 dBm. The LO generator designed in BiCMOS-55X process includes a PLL and a sextupler: the VCO reaches a 10–11.5 GHz TR and a simulated phase noise of −136 dBc/Hz at 1 MHz offset with a DC power of 380 mW, while the sextupler has a bandwidth from 60 to 78 GHz with a DC power of 67 mW.

REFERENCES

[1] M. de Kok, A. B. Smolders and U. Johannsen, "A Review of Design and Integration Technologies for D-Band Antennas," in IEEE Open Journal of Antennas and Propagation, vol. 2, pp. 746–758, 2021, doi: 10.1109/OJAP.2021.3089052.

[2] P. Chevalier et al, "B55X: A SHIFT in STMicroelectronics BiCMOS Technologies", 2024 19th European

Microwave Integrated Circuits Conference (EuMIC), Paris, France, 2024.

[3] I. Petricli, D. Riccardi and A. Mazzanti, "D-Band SiGe BiCMOS Power Amplifier With 16.8dBm P-dB and 17.1% PAE Enhanced by Current-Clamping in Multiple Common-Base Stages," in IEEE Microwave and Wireless Components Letters, vol. 31, no. 3, pp. 288–291, March 2021, doi: 10.1109/LMWC.2021.3049458.

[4] E. Rahimi, J. Zhao, F. Svelto and A. Mazzanti, "High-Efficiency SiGe-BiCMOS E-Band Power Amplifiers Exploiting Current Clamping in the Common-Base Stage," in IEEE Journal of Solid-State Circuits, vol. 54, no. 8, pp. 2175–2185, Aug. 2019, doi: 10.1109/JSSC.2019.2909475.

[5] A. Bilato, I. Petricli and A. Mazzanti, "SiGe BiCMOS D-Band Heterodyne Power Mixer With Back-Off Efficiency Enhanced by Current Clamping," in IEEE Solid-State Circuits Letters, vol. 7, pp. 2–5, 2024, doi: 10.1109/LSSC.2023.3332766.

[6] G. Amendola, L. Boccia, F. Centurelli, P. Chevalier, A. Fonte, C. Mustacchio, A. Pallotta, P. Tommasino, A. Traversa, A. Trifiletti, 'Compact E-Band I/Q Receiver in SiGe BiCMOS for 5G Backhauling Applications,' IEEE Transactions on Circuits and Systems II: Express Briefs, vol. 68, no. 9, Sept. 2021.

[7] G. De Filippi, L. Piotto, A. Bilato, and A. Mazzanti, 'A SiGe BiCMOS D-band LNA with gain boosted by local feedback in common-emitter transistors,' In 2023 IEEE Radio Frequency Integrated Circuits Symposium (RFIC), pp. 133–136, San Diego, CA, USA, 11-13 June 2023.

[8] S. Yoo et al., "23.4 An 82fsrms-Jitter and 22.5mW-Power, 102GHz W-Band PLL Using a Power-Gating Injection-Locked Frequency-Multiplier-Based Phase Detector in 65nm CMOS," in 2021 IEEE International Solid-State Circuits Conference (ISSCC), San Francisco, CA, USA, 2021, pp. 330–332, doi: 10.1109/ISSCC42613.2021.9365956.

[9] M. Mercandelli, L. Grimaldi, L. Bertulessi, C. Samori, A. L. Lacaita and S. Levantino, "A Background Calibration Technique to Control the Bandwidth of Digital PLLs," IEEE Journal of Solid-State Circuits, vol. 53, no. 11, pp. 3243–3255, Nov. 2018, doi: 10.1109/JSSC.2018.2866454.

[10] A. Franceschin, D. Riccardi and A. Mazzanti, "Ultra-Low Phase Noise X-Band BiCMOS VCOs Leveraging the Series Resonance," IEEE Journal of Solid-State Circuits, vol. 57, no. 12, pp. 3514–3526, Dec. 2022, doi: 10.1109/JSSC.2022.3202405.

[11] M. M. Pirbazari and A. Mazzanti, "E-Band Frequency Sextupler With >35 dB Harmonics Rejection Over 20 GHz Bandwidth in 55 nm BiCMOS," IEEE Journal of Solid-State Circuits, vol. 57, no. 7, pp. 2155–2166, July 2022, doi: 10.1109/JSSC.2022.3146730.

Alessandro Fonte received M.S. and Ph.D. degrees in electronic engineering from the University of Pisa, Italy, in 2005 and 2008, respectively. Since 2011, he has been with the SIAE MICROELETTRONICA S.p.A. (Milan, Italy), where he is a senior microwave engineer and member of the technical staff. His current research interests include the design of micro- and millimeter-wave system-on-chip front-ends in silicon and III-V semiconductor technologies for point-to-point radio link applications. He has authored several papers in peer-reviewed international journals and conference proceedings and two book chapters. He is the SIAE contact person for European and ESA projects.

Stefano Moscato (S'12) was born in Pavia, Italy, in 1988. He received the Ph.D. degree in electronics engineering from the University of Pavia, Italy, in 2016. He was a visiting Ph.D. student at Georgia Tech, Atlanta, GA, USA, in early 2015. He became part of the R&D microwave group of SIAE MICROELETTRONICA in May 2017. His research activities have been focused on RF-to-mmWave passive components. Since September 2022, Dr. Moscato has been the coordinator of the 1337 R&D group devoted to the design and validation of mmWave passive components, antennas, and sub-systems. He is involved in innovation programs and founded research for microwave backhauling, O-RAN equipment, and space-oriented assemblies. He was a recipient of an IEEE MTT-S undergraduate/pre-graduate scholarship in 2012. He is the author of more than 50 papers on international journals and conferences. He has been the Chair of the IEEE Student Branch, University of Pavia, from 2013 to 2016.

Riccardo Moro received the M.S. and the Ph.D. degrees in electronic engineering in 2010 and 2013, respectively, from the University of Pavia, Italy. In 2014, he was post-doc researcher at the same university. His main activity was focused on the design of planar components and antennas for radiofrequency applications. From 2015 to 2022, he worked in both consulting firm and telecommunication companies where he gained experience in technical management and analysis of microwave systems. In 2022, he joined SIAE MICROELETTRONICA as a senior RF engineer and he is involved in products development as well as in R&D funded projects.

Andrea Pallotta received the M.S. degree in electronic and telecommunication engineering from the University of Ancona, Italy, in 1990. After about one year period within the University of Ancona as a grant researcher, from 1991 to 1999, he has been with the Optoelectronic R&D Lab of Italtel, involved in many national and European projects as a senior designer. From 1998 to 2000, he was with the SDH Cross-Connect competence center of Siemens as responsible for the Advanced Development Department. He joined the ST end of 2000 as Design and Project Manager by leading the Optoelectronic and High Frequency Lab. His work has included the design and development of optoelectronic modules, GaAs MMIC, Bipolar, BiCMOS, and BCD-based silicon RFICs for high-speed data link, RF power amplifiers, and RF transceivers. He is leading, for more than 20 years, both product development and coordination activities of several European R&D funded projects in the field of the optoelectronic and RF telecommunication systems. He is the author and co-author of 19 technical papers and holds 15 patents.

Andrea Mazzanti received the Laurea and Ph.D. degrees in electrical engineering from the University of Modena and Reggio Emilia, Modena, Italy, in 2001 and 2005, respectively. During the summer of 2003, he was with Agere Systems, Allentown, PA, as an Intern. From 2006 to 2009, he was an assistant professor with the University of Modena and Reggio Emilia. In January 2010, he joined the University di Pavia, where he is currently a full professor of electronics and the scientific director of a joint lab between University and STMicroelectronics. His main research interests cover device modeling and IC design for high-speed communications, RF, and millimeter-wave systems. Dr. Mazzanti was a member of the Technical Program Committee of the IEEE Custom Integrated Circuit Conference (CICC), European Solid State Circuits Conference, and International Solid State Circuits Conference (ISSCC). He was an associate editor for the Transactions on Circuits and Systems and IEEE Solid-State Circuits Letters and a Guest Editor for a number of special issues of the Journal of Solid State Circuits. He currently serves as a Distinguished Lecturer of the IEEE Solid-State Circuits Society.

Andrea Bilato received the B.S. degree in information engineering, M.S. degree in electronic engineering, and the Ph.D. degree from the University of Padua, Padua, Italy, in 2015, 2017, and 2021, respectively. During his Ph.D., from 2019 to 2020, he was an intern with Infineon Technologies AG, Munich, Germany, where he worked on transceiver front-end circuits for radar systems operating beyond 100 GHz. Since 2021, he has been a post-doctoral researcher with the Department of Industrial and Information Engineering, University of Pavia, Pavia, Italy. His current research interests include the IC design of D-band transceiver building blocks for wireless high-speed communications.

Lorenzo Piotto was born in Padua, Italy, in 1994. He received the B.Sc. and M.Sc. degrees in electronics engineering from the University of Padua, Padua, Italy, in 2016 and 2018, respectively. From 2019 to 2020, he was with Infineon Technologies Austria, Villach, Austria, where he worked on the design and verification of mmWave blocks for wireless communication. In 2021, he joined the University of Pavia where he is working toward his Ph.D. His current research interest is in the design of building blocks for high-data-rate backhaul communications operating in D-band.

Guglielmo De Filippi was born in Milan, Italy, in 1993. He received the B.Sc. and M.Sc. degrees in electronics engineering from Politecnico di Milano, Italy, in 2017 and 2020, respectively, and the Ph.D. degree in microelectronics from the University of Pavia, Italy, in 2024.

In 2019, he was a thesis student with the ARP Laboratory, Politecnico di Milano, where he focused on mmWave building blocks for E-band wireless communications. In 2020, he joined the AIC Laboratory with the University of Pavia, Pavia, Italy. His current research interest is in D-band building blocks for high-data-rate wireless backhaul communications and radars.

Francesco Centurelli was born in Roma, in 1971. He received the laurea degree (cum laude) and the Ph.D. degree in electronic engineering from the University of Roma "La Sapienza", Roma, Italy, in 1995 and 2000, respectively. In 2006, he became an assistant professor with the DIET Department of the University of Roma La Sapienza. His research interests were initially focused on system-level analysis and design of clock recovery circuits and high-speed analog integrated circuits, and now concern the design of analog-to-digital converters and very-low-voltage circuits for analog and RF applications. He has authored or co-authored more than 100 papers on international journals and refereed conferences, and has been also involved in R&D activities held in collaboration between Università "La Sapienza" and some industrial partners.

Pietro Monsurrò was born in Rome, Italy. He received the Ph.D. degree in electronic engineering from the Department of Electronic Engineering, University of Roma "La Sapienza," Italy, in 2008. He is currently an associate professor with the Department of Information Engineering, Electronics and Telecommunications of the University of Roma "La Sapienza," Italy. His current research interests include low-voltage low-power analog and mixed-signal circuits in advanced CMOS processes, analog filter design in CMOS and BiCMOS processes and filter synthesis techniques, behavioral models of mixed-signal and RF systems, digital background calibration techniques for pipeline, and time-interleaved analog-to-digital converters, including advanced time-interleaved architectures, model identification, and adaptive algorithms.

Hassan Sadeghi Chameh received the master's degree in communication engineering with the Cardiff University and is currently studying the Ph.D. degree in electronic engineering with the University of Roma "La Sapienza." His research interests are in the design of integrated circuits for E-band and D-band transceivers.

Pasquale Tommasino received the master's degree in electronic engineering and the Ph.D. degree from the University of Roma "La Sapienza" in 1992 and 1999, respectively. Since 1995, he has been with the Electronic Engineering Department (now Department of Information Engineering, Electronics and Telecommunications) of the University of Roma "La Sapienza," where since 2007 he has been an assistant professor. His research interests are mainly oriented to the design of RF and microwave broadband circuits. Presently, he is mainly engaged in the design of integrated circuits for E-band and D-band transceivers.

Alessandro Trifiletti was born in Rome, Italy, on October 4, 1959. In 1991, he joined the Electronic Engineering Department of "La Sapienza" University in Rome as a research assistant, where he was involved in research activities dealing with analogue, RF, and microwave IC's design. In 2001, he became an assistant professor and in 2005, he got the position of Associate Professor and in 2019 the position of Full Professor at the Engineering Faculty of the same university. Prof. Trifiletti has worked in the field of microelectronics, both from the point of view of design methodologies and circuit topologies. On these subjects, Prof. Trifiletti has (co-) authored over 210 publications, of which about 80 published on international journals, the others published on the proceedings of major international conferences (a large part of these sponsored by the IEEE). He is presently a reviewer for some IEE and IEEE reviews, among them: IEEE Transactions on Microwave Theory and Techniques, IEEE Transactions on Circuit and Systems (parts I and II), IEE Proceedings on Circuits, Devices and Systems, and IEE Electronic letters. In the last 20 years, he has been engaged in the coordination of research teams from DIET (previously DIE) in the framework of national and international programs, involving both industrial and academic partners. From an industrial perspective, Prof. Trifiletti's expertise covers topics about analogue and RF microelectronics, Radar and Electronic Support Measures (ESM) systems, high-speed communication systems, security issues in cryptographic algorithms implementation, and embedded system design.

Daniele Lodi Rizzini was born in Casalmaggiore, Italy, in 1999. He received the B.Sc. and M.Sc. degrees in electronics engineering from the Politecnico di Milano, Milan, Italy, in 2021 and 2023, respectively, where he is currently a research assistant.

His current research interests include low jitter fractional-N frequency synthesizers and frequency multiplication circuits used in millimeter-wave transceivers.

Francesco Tesolin received the M.S. and the Ph.D. degree (cum laude) in electrical engineering from the Politecnico di Milano, Milan, Italy, in 2019 and 2024, respectively, where he is currently working as a post-doctoral researcher.

His current research interests include design of digitally controlled oscillators (DCOs) and low-jitter digitally intensive fractional-N frequency synthesizers for communication systems and automotive radars. He was a co-recipient of the Jan Van Vessem Award for Outstanding European Paper of the 2023 International Solid State Circuits Conference.

Simone M. Dartizio was born in Milan, Italy, in 1995. He received the M.Sc. and Ph.D. degrees (cum laude) in electrical engineering from the Politecnico di Milano, Milan, Italy, in 2019 and 2023, respectively. From 2023, he has been a post-doctoral researcher with the Politecnico di Milano. His research interests include low-jitter frequency synthesis based on digitally intensive phase-locked loops and digital calibrations for mixed-signal integrated circuits. He was the recipient of the ISSCC 2023 Jan Van Vessem award for outstanding European paper and the 2023 SSCS Predoctoral Achievement award.

Salvatore Levantino received the Laurea degree (cum laude) and the Ph.D. degree in electrical engineering from the Politecnico di Milano, Milan, Italy, in 1998 and 2001, respectively. From 2000 to 2002, he was a consultant with Bell Labs, Lucent Technologies, Murray Hill, NJ, USA. Since 2005, he has been with the Politecnico di Milano, where he is currently a full professor of electrical engineering. He has co-authored over 180 scientific articles and the book "Integrated Frequency Synthesizers for Wireless Systems" (Cambridge University Press, 2007). He holds six patents. His research interests include wireless transceivers, frequency synthesizers, data converters, and power converters. He was a member of the Technical Program Committee of the International Solid-State Circuits Conference (ISSCC), the Radio Frequency Integrated Circuits (RFIC) Symposium, and the European Solid-State Circuits Conference (ESSCIRC). He was a co-recipient of the ISSCC 2023 Jan Van Vessem award for outstanding European paper and the Best Paper Award of the 2018 International Symposium on Circuits and Systems (ISCAS). He was a Guest Editor for two issues of the *IEEE Journal of Solid-State Circuits* and an associate editor for the IEEE Transactions on Circuits and Systems—I: Regular Papers and the IEEE Transactions on Circuits and Systems—II: Express Briefs. He was an IEEE SSCS Distinguished Lecturer.

Sub-THz Transceiver Design for Future Generation Mobile Communications

G. Mangraviti, C. Dehos, V. Puyal, O. Richard, G. Yaakoubi khbiza, F. Foglia Manzillo, X. V. L. Nguyen, F. Filice, E. Noccetti, P. L. Hellier, F. Podevin, S. Bourdel, A. Ruffino, K.-S. Choi, B. Abdelaziz Abdelmagid, M. Eleraky, H. Wang, B. Ardouin, T. K. Johansen, L. Fanori, Q. Huang, E. Schlaffer, P. Wambacq, D. Morche, and B. Debaillie

Abstract—**This paper outlines the design considerations necessary to realize an innovative transceiver prototype for future-generation mobile communication, adeptly harnessing the spectrum beyond 100 GHz. The primary innovations and challenges reside in maximizing the efficacy of BiCMOS and indium phosphide technologies, advanced radio frequency (RF) packaging, and the design of high-performance D-band RF front-ends. The selection of RF-chip technologies and the integration of densely packed RF packaging are thoroughly defined and justified. Specifications for both transmitter and receiver systems are derived from a meticulous link budget analysis. These preliminary studies and decisions inform the forthcoming tape-outs in this project. The focus remains on developing key transceiver technologies to drive the next generation of mobile communications, surpassing the capabilities of 5G. This includes enhancing data rates, power efficiency, integration density, and minimizing footprint.**

Index Terms—**6G mobile communication, antenna-in-package, BiMOS integrated circuits, InP, integrated circuit packaging, millimeter wave integrated, mobile communication, phased arrays, receivers, transmitters**

This work is supported by the Chips Joint Undertaking (Chips JU) under grant agreement number 101096256 (project SHIFT). The Chips JU receives support from the European Union's Horizon Europe research and innovation program and the National Authorities.

G. Mangraviti, F. Filice, E. Noccetti, P. Wambacq, and B. Debaillie are with imec, Belgium. P. Wambacq is also with Vrij Universiteit Brussel, Belgium, and E. Nocetti is also with Università di Pisa, Italy. C. Dehos, V. Puyal, O. Richard, G. Yaakoubi Khbiza, F. Foglia Manzillo, X. V. L. Nguyen, and D. Morche are with CEA-Leti, France. P. L. Hellier, F. Podevin, and S. Bourdel are with Université Grenoble Alpes, France. A. Ruffino, K.-S. Choi, B. Abdelaziz Abdelmagid, M. Eleraky, and H. Wang are with ETH Zürich, Switzerland. B. Ardouin is with III-V Lab. T. K. Johansen is with Danmarks Tekniske Universitet, Denmark. L. Fanori and Q. Huang are with Advanced Circuit Pursuit (ACP), Switzerland. E. Schlaffer is with AT&S, Austria. Contact: G. Mangraviti, Giovanni.Mangraviti@imec.be

I. INTRODUCTION

According to market analysts, the proliferation of smart devices is expected to continue, contingent upon technological advancements. Devices like augmented reality glasses require high bandwidth connectivity, efficient operation, and dense integration. Utilizing higher frequency bands above 100 GHz holds significant potential to meet these emerging system requirements and also presents severe technical challenges. Within the SHIFT (Sustainable technologies enabling Future Telecom applications) project, a Chips JU initiative focused on advancing sustainable technologies to enable future telecom applications, we are developing a "Future-G transceiver" prototype. This prototype leverages the D-band radio spectrum to achieve the demanding requirements of speed and bandwidth.

Currently, such operating frequencies are restricted to wireline communication, while long-range wireless links beyond 100 GHz are only established for static applications (i.e. back/front-hauling) rather than for mobile applications. At these frequencies, hardware validation necessitates most advanced measurement and lab infrastructure. Most developed demonstrators are still using horn antennas, which are neither integrated nor compact.

In addition to high frequency, technology must also support efficiency. Recent advancements in BiCMOS systems, for instance, demonstrate data rates of tens of Gb/s in the frequency bands above 100 GHz [1][2][3]. However, the limited output power of this technology restricts the link distance to a few centimeters. Conversely, III-V technologies, such as indium phosphide (InP) from III-V Lab, allow for efficient and high radio frequency (RF) performance beyond 100 GHz [4][5].

We address efficiency by combining (i) ST BiCMOS B55X technology [6], known for its state-of-the-art performance in RF and mixed signal on silicon, with (ii) III-V-Lab's InP technology, which provides superior amplification efficiency and power delivery beyond 100 GHz. A system study illustrating the benefits of combining silicon transmitters (TXs) with InP power amplifiers (PA) is shown in Figure 1 [7]. This study demonstrates that for an N-path phased-array TX with a total equivalent isotropic radiated power (EIRP) of 40 dBm, the combination of a silicon-based system with InP

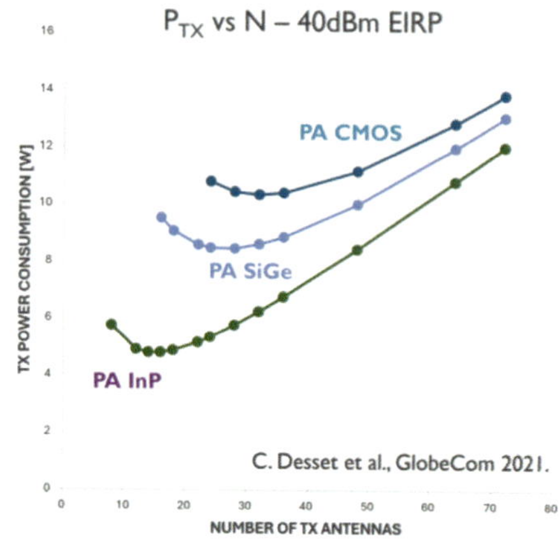

Fig. 1. Study in [7]: benefit of combining InP PAs with Silicon-based phased-array sub-THz transmitter.

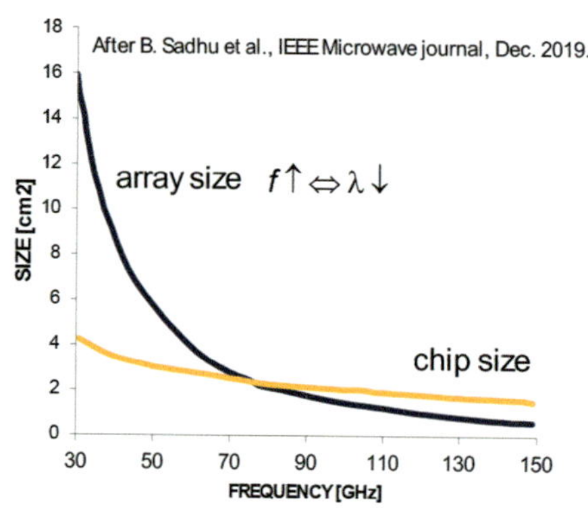

Fig. 2. Array size versus chip size, after [8].

PAs reduces both power consumption and the optimal number (*N*) of antennas.

Beyond high operating frequency and efficiency, RF packaging poses significant challenges in terms of signal losscs and size. At frequencies beyond 100 GHz, antenna gain decreases while the connectivity loss between on-chip amplifiers and antennas increases. Moreover, the size of the phased-array becomes comparable to the chip size, as shown in Figure 2, after the study in [8].

Our work addresses key innovations and challenges in developing integrated solutions for future high-speed mobile communication. These include

TABLE 1.
SCENARIO KEY PERFORMANCE INDICATORS.

Demonstrator KPI	**Scenario's targets**	**Demonstrator's targets**
Data rate	2 channels – 60 Gsps	1 channels – 30 Gsps
Bandwidth	2 × 5GHz in dual polarization	5 GHz in single polarization
Range	100 m	5m
Tx EIRP	24dBm	24dBm
Form factor	12 antennas	4 antennas
Frequency of operation	~140 GHz	~140 GHz
Modulation order	64 QAM	16 QAM
Transmit power and efficiency	18 dBm OCP1 dB 20% PAE max	16 dBm OCP1dB 10% PAE max

the optimal selection and utilization of chip technologies such as BiCMOS and InP, the design of high-performance and efficient D-band (140 GHz, wideband) TX and receiver (RX) front-ends, and advanced RF packaging. Starting with a high-data-rate link budget scenario in Section II, initial specifications for the TX and RX front-ends are derived in Section III. Various aspects of the implementation are detailed in Section IV, and conclusions are presented in Section V.

II. SCENARIO OF HIGH DATA-RATE LINK

The scenario considered in this demonstrator involves establishing a high-data-rate communication link between a base station (BS) and user equipment (UE) in environments such as airports or shopping malls. Table I outlines the key performance indicators (KPIs) for the proposed scenario. Note that while the system is designed to meet the scenario's targets, the requirements have been moderated for the final demonstrator. This approach is intentionally adopted to mitigate risks and ensure a functional demonstrator within the project SHIFT's timeframe.

The link budget assumes a line-of-sight (LoS) link between a BS and a mobile UE, aiming to cover distances up to 100 m. The system aims to achieve a raw data rate of up to 60 Gbit/s with dual polarization and 30 Gbit/s with single polarization (maximized in downstream, can be smaller in upstream). Given a single stream of 30 Gbit/s and a raw spectral efficiency of 6 bit/s/Hz using 64 quadrature-amplitude modulation (QAM), the required channel bandwidth is 5 GHz.

The EIRP and the total radiated power (TRP), crucial specifications for the link budget, are drawn from the ETSI 5G FR2 standard [9]. That because D-band communication is not yet standardized. In our scenario, the baseband TX is referred to the power class 1, imposing limits of 55 dBm for EIRP and 35 dBm for TRP. Conversely, the handheld UE TX falls under power class 3, with EIRP and TRP limited to 43 and 23 dBm, respectively.

The link budgets for the downstream reception at the user equipment (UE RX) and upstream transmission from the user equipment (UE TX) are illustrated in Figures 3 and 4. In these figures, we utilize the modulation-and-coding scheme (MCS) of the WiGiG 60-GHz standard [10] chosen for its ability to accommodate a much wider bandwidth compared to current 5G FR2 standards. The TX output 1-dB compression point (OP1dB) is deliberately specified, and the number of antennas on the TX and RX sides is carefully adjusted

MCS (derived from WiGig)	0	9	13	17	
Constellation	D-BPSK	QPSK	16QAM	64QAM	
LDPC code rate	1/2	13/16	13/16	13/16	
Spreading/repetition	32	1	1	1	
raw data rate	5000	10000	20000	30000	Mbps
Frequency					
Carrier frequency	140000				MHz
wavelength (lambda)	0.0021				m
bandwidth	5000				MHz
roll-off	0				
TX (basestation)					
Tx power per PA (output P_1dB)	12.00	14.00	16.00	18.00	dBm
Number of antennas per sub-array (TX)	1.00	1.00	1.00	1.00	
Number of sub-arrays (TX)	128.00	128.00	128.00	128.00	
Lens gain	0.00	0.00	0.00	0.00	dB
Antenna element gain	5.00	5.00	5.00	5.00	dBi
Total array gain (20log(Ntx) + ...)	47.14	47.14	47.14	47.14	dB
TX interconnect loss (<0 dB)	-4.00	-4.00	-4.00	-4.00	dB
Back-off from P_1dB (>= 0)	0.00	2.00	4.00	6.00	dB
Total EIRP	55.14	55.14	55.14	55.14	dBm
TRP	29.07	29.07	29.07	29.07	dBm
Channel					
LOS loss	-141.35	-121.71	-114.79	-105.88	dB
Atmospheric attenuation factor (LOS)	**0.000**				dB/m
Atmospheric attenuation (with range)	0.00	0.00	0.00	0.00	dB
RX (UE)					
Number of antennas per sub-arrays (RX)	1.00	1.00	1.00	1.00	
Number of sub-arrays (RX)	12.00	12.00	12.00	12.00	
Lens gain	0.00	0.00	0.00	0.00	dB
Antenna element gain	5.00	5.00	5.00	5.00	dBi
Total array gain (10log(Nrx) + ...)	15.79	15.79	15.79	15.79	dB
RX interconnect loss (<0dB)	-4.00	-4.00	-4.00	-4.00	
Received power including array gain	-74.42	-54.78	-47.85	-38.94	dBm
Thermal noise	-77.01	-77.01	-77.01	-77.01	dBm
Noise figure	9.00	9.00	9.00	9.00	dB
Implementation loss (Analog and Digital)	-3.00	-3.00	-3.00	-3.00	dB
EVM from non-additive effects (phase noise, ...)	-23.00	-23.00	-23.00	-23.00	dB
Link margin	2	2	2	2	dB
Input referred noise (including EVM)	-63.01	-62.87	-62.35	-59.43	dBm
SNR, Range, Bit rate					
Required SNR (coded BER = 3e-7)	-11.40	8.10	0.00	0.00	dB
Achieved SNR	-11.41	8.09	14.50	20.49	dB
SNR difference - link margin	-0.01	-0.01	14.50	20.49	dB
Max range	1992.73	207.71	93.59	33.55	m

Fig. 3. Link budget for UE RX downstream.

MCS (derived from WiGig)	0	9	13	17	
Constellation	D-BPSK	QPSK	16QAM	64QAM	
LDPC code rate	1/2	13/16	13/16	13/16	
Spreading/repetition	32	1	1	1	
raw data rate	5000	10000	20000	30000	Mbps
Frequency					
Carrier frequency	140000				MHz
wavelength (lambda)	0.0021				m
bandwidth	5000				MHz
roll-off	0				
TX (UE)					
Tx power per PA (output P_1dB)	14.00	16.00	18.00	18.00	dBm
Number of antennas per sub-array (TX)	1.00	1.00	1.00	1.00	
Number of sub-arrays (TX)	16.00	16.00	16.00	16.00	
Lens gain	0.00	0.00	0.00	0.00	dB
Antenna element gain	5.00	5.00	5.00	5.00	dBi
Total array gain (20log(Ntx) + ...)	29.08	29.08	29.08	29.08	dB
TX interconnect loss (<0 dB)	-4.00	-4.00	-4.00	-4.00	dB
Back-off from P_1dB (>= 0)	0.00	2.00	4.00	6.00	dB
Total EIRP	39.08	39.08	39.08	37.08	dBm
TRP	22.04	22.04	22.04	20.04	dBm
Channel					
LOS loss (Friis)	-135.57	-115.93	-109.01	-98.10	dB
Atmospheric attenuation factor (LOS)	**0.000**				dB/m
Atmospheric attenuation (with range)	0.00	0.00	0.00	0.00	dB
RX (basestation)					
Number of antennas per sub-arrays (RX)	1.00	1.00	1.00	1.00	
Number of sub-arrays (RX)	128.00	128.00	128.00	128.00	
Lens gain	0.00	0.00	0.00	0.00	dB
Antenna element gain	5.00	5.00	5.00	5.00	dBi
Total array gain (10log(Nrx) + ...)	26.07	26.07	26.07	26.07	dB
RX interconnect loss (<0dB)	-4.00	-4.00	-4.00	-4.00	
Received power including array gain	-74.42	-54.78	-47.85	-38.94	dBm
Thermal noise	-77.01	-77.01	-77.01	-77.01	dBm
Noise figure	9.00	9.00	9.00	9.00	dB
Implementation loss (Analog and Digital)	-3.00	-3.00	-3.00	-3.00	dB
EVM from non-additive effects (phase noise, ...)	-23.00	-23.00	-23.00	-23.00	dB
Link margin	2	2	2	2	dB
Input referred noise (including EVM)	-63.01	-62.87	-62.35	-59.43	dBm
SNR, Range, Bit rate					
Required SNR (coded BER = 3e-7)	0.00	0.00	0.00	0.00	dB
Achieved SNR	-11.41	8.09	14.50	20.49	dB
SNR difference - link margin	-11.41	8.09	14.50	20.49	dB
Max range	1024.03	106.74	48.10	13.70	m

Fig. 4. Link budget for UE TX upstream.

to achieve a 100-m link distance without exceeding the EIRP or TRP limits. For the UE RX downstream, a distance of 100 m is only achievable for modulation schemes

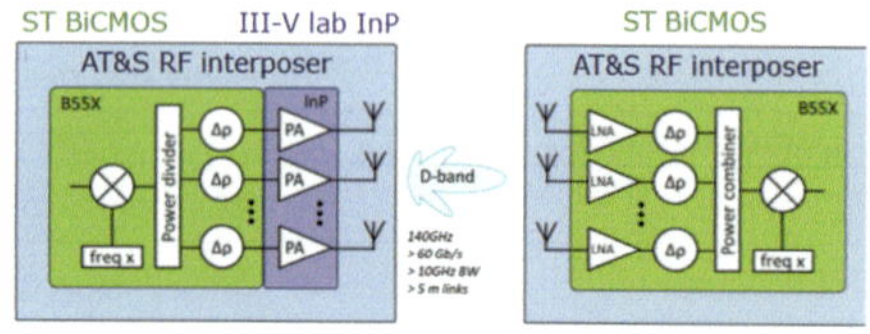

Fig. 5. Transmitter front-end integrating BiCMOS and InP chips within an advanced RF packaging (on the left) and receiver front-end integrating a BiCMOS chip within an advanced RF packaging (on the right).

up to 16 QAM, with a raw data rate up to 20 Gb/s, but not for 64 QAM, which offers a raw data rate of 30 Gb/s. Similarly, for the UE TX upstream, a distance of 100 m is only feasible for modulation schemes up to QPSK (Quadrature Phase Shift Keying), providing a raw data rate up to 10 Gb/s. It is important to note that these specifications are intended for the system (and subcircuits) design purposes, and the requirements for the final demonstrator will be relaxed to mitigate risks.

III. SPECIFICATIONS

The integration concept of TX and RX envisioned in this project is shown in Figure 5. The RX front-end has a chip in ST BiCMOS B55X technology, embedded within an advanced RF packaging by AT&S [11]. The TX front-end comprises a chip in B55X and last stages of PAs in III-V-Lab InP technology. To showcase beamforming, both TX and RX implement phase-shifting array on chip.

Two 5-GHz channel bandwidths are targeted: from 143.5 to 148.5 GHz and from 151.5 to 156.5 GHz. These are a part of the four bandwidths recommended by [12], as shown in Figure 6. Accordingly, the TX and the RX front-ends are specified to work from 143.5 to 156.5 GHz.

A. *Transmitter*

The primary specifications for the transmitter (TX) front-end, as determined from

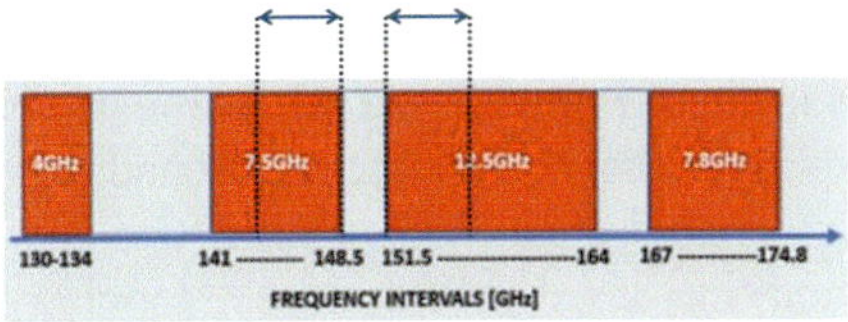

Fig. 6. Channel bandwidths recommended by [12], and targets for this project: 143.5–148.5 GHz and 151.5–156.5 GHz.

the link budget (Figure 4), include output power, bandwidth, and signal-to-noise ratio (SNR). Additionally, a key focus is placed on achieving the overall efficiency. The output power is currently set at a 1-dB compression point of 18 dBm. However, ongoing studies are exploring the trade-off between this high power requirement and the efficiency of InP, which may result in a potential downgrade of this specification to 15 dBm. Ensuring an SNR well above 23 dB for a coded 64-QAM modulation scheme is crucial. To achieve this, we have opted for an up-conversion transmitter (TX) topology based on a wideband intermediate frequency (IF) instead of a zero-IF one. This decision helps mitigate SNR degradation caused by baseband direct current (DC) offset and in-phase/quadrature (I/Q) imbalance.

B. Receiver

We propose a four-path RX architecture, shown in Figure 7, as a trade-off between ease of integration in phased arrays, silicon area, power consumption, and heat dissipation. The considered application requires four RX antenna paths, synchronized by their common frequency reference input. The noise figure of the RX is reduced by the shrinking of the antenna interconnection length due to 3D packaging, and innovative LO phase shifting, instead of lossy RF phase shifting (vector modulator, switched delay cells, etc.). We expect to save power consumption and area while improving performance. Here we consider zero-IF direct conversion mixing over a wide bandwidth (2 × 2.5 GHz).

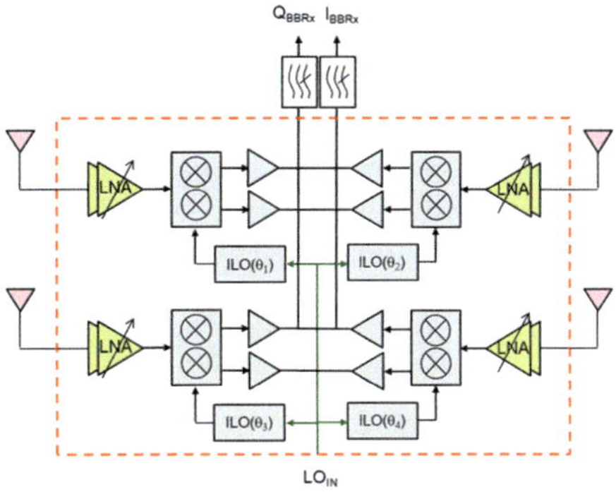

Fig. 7. Proposed receiver architecture.

TABLE II.
MAIN RECEIVER SPECIFICATIONS.

Receiver	Gain: 10–57 dB, NF: 9 dB, BW: 142.5–157.5 GHz
LNA	Gain: 18/10 dB, NF: 7.5 dB, IIP3: −15/−7 dBm
I/Q mixer	Gain: 3/−9 dB, NF: 20 dB, IIP3: 3/11 dBm, 3° phase imbalance
LOin	Frequency: 36.5/38.5 GHz, Phase noise: −107 dBc/Hz@1 MHz & −137dBc/Hz noise floor
Phase shifters	2 + 4 bits resolution

From the lowest sensitivity level to the maximum input power, the RX shall cope with a power dynamic range of 47 dB, from −67 to −20 dBm. Part of it (30 dB) shall be handled by the base band variable gain amplifier (VGA). Moreover, low gain modes shall be considered for low-noise amplifier (LNA) and mixer to de-saturate the RX. A summary of specifications is shown in Table II.

IV. IMPLEMENTATION

A. Transmitter Front-End in BiCMOS

The TX architecture proposed in the SHIFT project, shown in Figure 5, incorporates a wideband IF system, followed by a BiCMOS phased-array TX front-end, used to drive InP PAs. An RF interposer will be used to host the BiCMOS/InP chips and to integrate antennas and interconnections.

In the BiCMOS TX front-end, the IF signal is up-converted to the D-band frequencies using a mixer. The local-oscillator (LO) signal is generated at a lower frequency and then a frequency multiplier is used to drive the mixer LO port. To avoid the severe losses of passive mixers, the TX up-conversion mixer topology is Gilbert cell-based. The output 1-dB compression point of the Gilbert cell block must be carefully designed to make sure that the mixer does not introduce severe compression to the chain under the large input power level required to saturate the PAs. A wideband matching at the IF port is achieved by either using a resistive-feedback inverter or by adding a resistive termination at the gate of the IF transistors to reduce the input impedance. The LO power level required to achieve acceptable linearity and gain is determined so that saturated LO drivers deliver this power level to the LO transistors with reasonable efficiency.

Following the mixer, power dividers are used to drive the phase shifters. For power splitters, the Wilkinson topology offers scalable power splitting with good isolation between the array elements. Wilkinson power splitters can be implemented based on transmission lines, lumped components, or transformers. Although transmission-line-based Wilkinson dividers generally have low loss, wide bandwidth, and are simple to implement, they are usually not adopted on chip at lower RF/mmWave frequencies (e.g., 30 GHz) due to the large area they require ($\lambda/4 \sim 2.5$ mm at 30 GHz). However, at 150 GHz, $\lambda/4$ drops to 500 μm or even lower length; therefore, transmission-line-based Wilkinson power splitters can be feasible at D-band with reasonable area. With such a transmission-line-based Wilkinson topology, matching and isolation > 12 dB can be expected and an insertion loss of around 0.5 dB (on top of the 3 dB fundamental loss) is expected across the whole D-band.

Finally, phase shifters precede the InP PAs. Phase shifters typically build active vector modulators or passive implementations. The latter mainly include reflective-type phase shifters, switched-type phase shifters, or switchable transmission lines. Reflective-type phase shifters require high-order loads (e.g., *CLC* loads) and/or cascading multiple units to achieve the 360° phase range. However, in the D-band, on-chip varactors often exhibit significant losses and offer limited capacitance tuning range. As a result, these limitations can lead to increased insertion loss or a reduction in the achievable phase range. Similarly, utilizing switched-type phase shifters or switchable transmission lines may necessitate a cascade of multiple stages with numerous switches, thereby increasing the insertion loss of the phase shifter to over 10 dB at D-band. To circumvent the insertion loss associated with passive implementations, the transmitter (TX) phase shifters will be based on a vector modulator architecture. VGAs using multiple cells of common-source stages should be implemented with different gain weights, connected in parallel. The power combination of the I/Q components by summing the outputs of the VGAs in the current domain, and the 0°/180° phase flipping is achieved by switching the terminals of the differential output. The phase shifters are planned with 4-bit accuracy (22.5° phase resolution) accordingly. Such a BiCMOS front-end will drive high-power InP PAs on the same interposer.

B. Power Amplifiers in InP

The InP DHBT (Double Heterojunction Bipolar Transistor) technology of III-V Lab [4] is natively optimized for high-speed analogue–digital electro-optical interfaces (for optical fiber applications): transistor's figures of merit are optimized for fast current switching, high voltage drive, and wideband operation. The first run of D-band PA in SHIFT is therefore essentially an evaluation of the achievable performance with respect to the state-of-the art, which will serve as a guideline for (i) improving transistor design and (ii) selecting circuit design options and techniques (power cell topology, matching methods, power combining options, etc.). The used 0.5-µm InP DHBT technology has an fT/fMAX ~ 380/520 GHz for a 0.5 × 5-µm2 device at Vce = 1.6V and Jc = 7 mA/µm2. Besides fT and fMAX, the breakdown voltage (BVceo = 4.2 V) and knee-voltage (Vk ~ 0.8 V) are also important metrics for PAs that are directly impacting output power (P_{sat}) and power-added efficiency (PAE). To estimate realistic expectations, an idealized parallel-tuned prototype PA has been initially considered, which showed an advantage for the class-AB operation in terms of PAE and P_{1dB} (compression point). Based on this, the common-base PA topology is shown in Figure 8.

The layout is based on thin-film microstrip lines for compactness and overall lowest interconnect loss. The design exploits two-finger 0.5 × 5-µm2 devices for the power stage and single-finger 0.5 × 7-µm2 device for the driver. The chosen topology draws inspiration from [5].

Initial simulation results indicate a gain G_{ss} = 8.5 dB, P_{1dB} = 12.2 dBm and PAE_{max} = 13.4% at 140 GHz. These findings

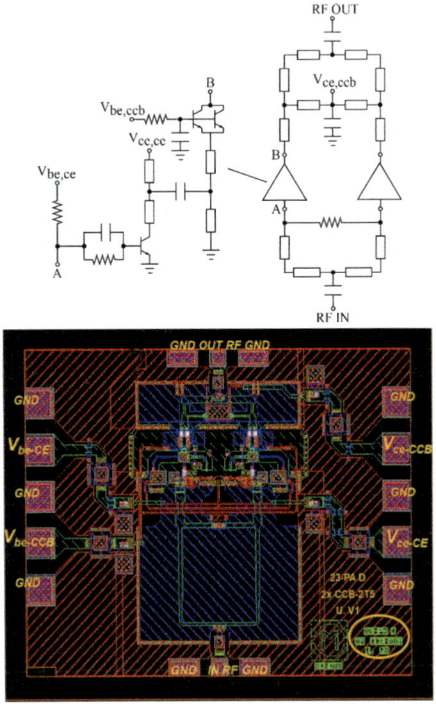

Fig. 8. Simplified schematic and layout of two-way combined power amplifier using common-base power stage with base capacitance and common-emitter driver.

underscore the necessity of employing 0.4 µm emitter width devices and optimized multi-finger DHBTs to enhance performance further. Currently, this endeavor is underway in preparation for the upcoming fabrication run. As anticipated, this initial design serves as a valuable reference point for future enhancements and aids in the selection of viable technology optimization strategies.

C. Wideband Intermediate Frequency

Generating a 5-GHz bandwidth signal at 150 GHz introduces a major challenge in designing both the baseband and the RF sections of the TX in a direct up-conversion architecture, especially if the error-vector-magnitude (EVM) layer must meet 64-QAM modulation requirements.

A preferable choice is to create a wide-band IF signal, based on a multi-carrier approach, which alleviates baseband complexity and ensures accurate LO quadrature generation with moderate power consumption.

The TX is tested with a 5-GHz bandwidth signal having IF placed between 4 and 5 GHz, which combines 4–8 channels together, each with maximum bandwidth of 100 MHz, EVM and quadrature accuracy commensurate with 64QAM modulation.

This solution adds flexibility for signal allocation within the bandwidth: in a non-contiguous scenario, the 4–8 channels can be distributed between 2 and 7 GHz, while in a contiguous scenario, the 400- to 800-MHz bandwidth supports multi-carrier performance evaluation.

D. Receiver Front-End in BiCMOS

In this paragraph, we will provide a brief overview of the implementation of the key blocks of the RX. The first critical building block is the LNA, with its most challenging specification being the noise figure. The schematic selected for the LNA is depicted in Figure 9. It features a classical two-stage cascode topology with input, inter-stage, and output matching networks based on transmission lines. Due to the limited maximum gain of bipolar transistors at 150 GHz, the LNA requires two cascode stages to achieve the specified overall gain.

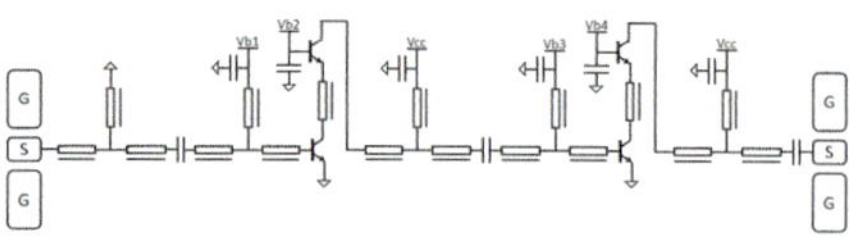

Fig. 9. LNA schematic.

The matching network topology ensures DC biasing through parallel stubs, and DC isolation between stages using link capacitors. Stubs are alternating current (AC) short-ended using custom decoupling capacitors. Matching networks are EM-simulated including capacitors to obtain reliable results.

Our first analysis of the topology shows that we can reach more than 20 dB of gain in the targeted bandwidth with a noise factor of 6.5 dB, leaving some room for the interconnection losses. Further investigations are ongoing to improve even more performance.

To generate the four quadrature phase-shifted LOs, we utilize the architecture depicted in Figure 10. Every path is fed by the same frequency reference that operates at one-fourth of the carrier frequency (36.5 and 38.5 GHz). The signal is multiplied by 2 in frequency, before rough phase shifting on 2 bits (0/45/90/135°) at 73 or 77 GHz. The signal is then again multiplied by 2 before the fine phase shifting, operated at the carrier frequency by an injection-locked phase shifter (ILPS). From Adler's theory [13], the output injection-locked-oscillator (ILO) phase shift varies between +/–90° within the locking range. We propose to use

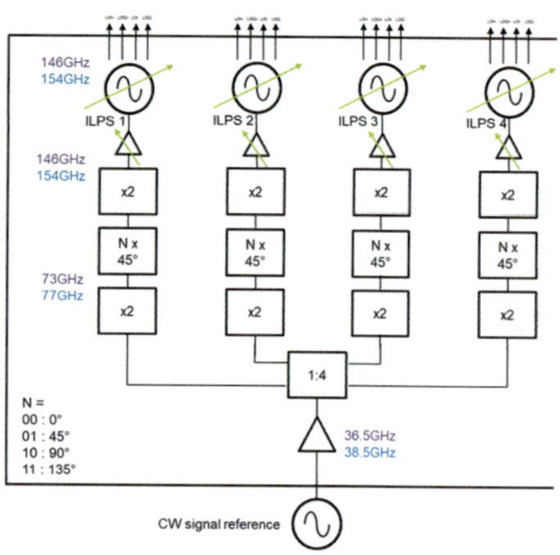

Fig. 10. Multi-LO frequency synthesis architecture.

only part of this range (+/–45°) for fine phase shifting, in order to guarantee locking margin and linear behavior close to zero. The phase shift can be obtained by playing on its natural pulsation through the variation of the ILO tank capacitance (switched capacitance and/or varactor tuning). It is also possible to adjust phase shift and locking range though the variation of the injection level [14]. The phase shifts to apply on the LOs are coded over 2 + 4 bits: 2 bits for the rough phase shifter, and 4 bits for the fine ILPS phase shifter. A 5-bit quantization is enough to generate beams with 3° angle steps and 0.25-dB inter-beam loss. However, 6 bits are intended for the calibration process.

The ILO implementation is based on a Colpitts oscillator, enhanced by incorporating a locking mechanism through an injected signal, as shown in Figure 11. This structure enables the oscillator to lock onto an external frequency source, thereby reducing phase noise and enhancing spectral purity.

The ILPS system utilizes the inherent characteristics of the Colpitts oscillator, such as its feedback network consisting of a combination of capacitors (C1, Cvar, and input Balun TR1), to establish a resonant circuit that determines the oscillation frequency, coupled with inductive degeneration (L deg) to achieve higher frequencies. By adjusting the feedback components, the oscillator can be finely tuned to the desired frequency (Figure 12), which is then locked by the injection of a stable reference signal.

The second critical block of the frequency synthesizer is a 2-bit 0/45/90/135° broadband phase shifter centered at 75 GHz to address the two usual frequencies (73 and 77 GHz). This 2-bit phase shifter using switched transmission line technique is shown in Figure 13. It is based on four transmission lines controlled by

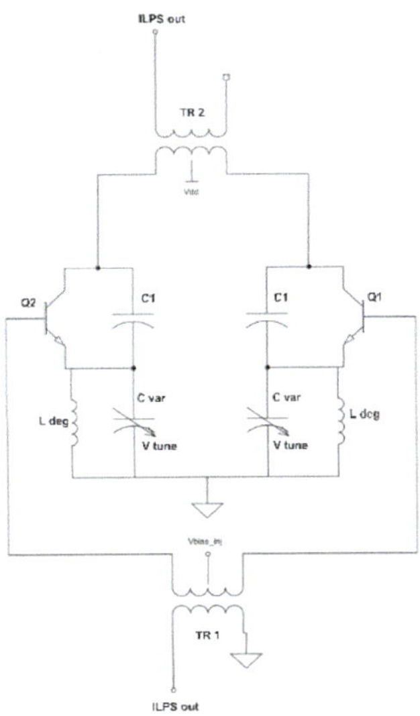

Fig. 11. Colpitt-based ILPS schematic.

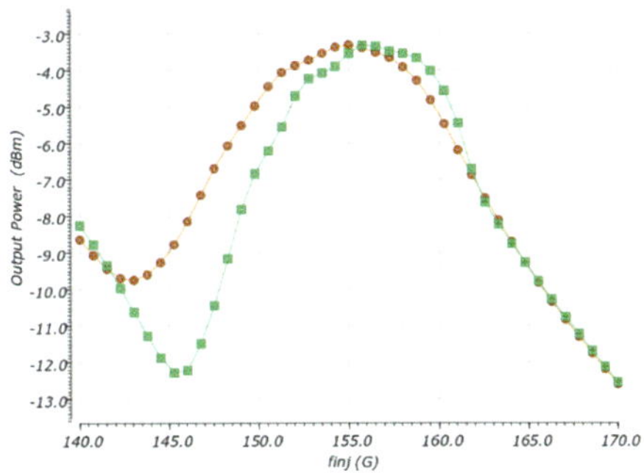

Fig. 12. Tuning of the locking bandwidth.

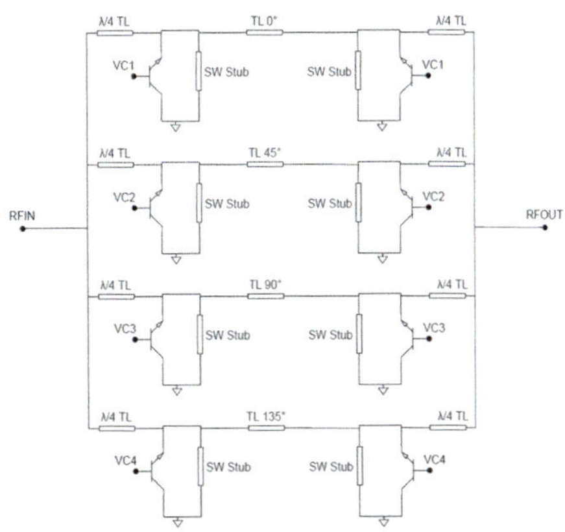

Fig. 13. Switched line phase shifter (SLPS) topology.

two quarter-wave shunt SP4T high-speed bipolar switches to take advantage of the benefits of the B55X technology. Each switched transmission line is sized to provide up to 135° of phase delay in 45° steps (with a maximum phase error of ±3.8° in the bandwidth). This passive architecture (SLPS: switched line phase shifter) was chosen to minimize insertion loss in our specific case (2 bits with a phase shift up to 135°). It is a trade-off between a classic usual passive topology such as SFPS (switched filter phase shifter [15]) that presents high RF losses and a more complex active vector modulator phase shifter type dedicated to 360° phase shift [16]. The power consumption is 7.4 mW. In terms of performance, the phase shifter cell can operate in 73–77 GHz bandwidth and the insertion loss is limited to 3.0 dB with higher than 17-dB return loss.

E. Advanced RF Packaging

The integration of chips and antennas poses a challenge owing to the high operating frequency and wide channel bandwidth. However, realizing the full potential of SiGe BiCMOS and InP advanced technologies demands the utilization of cutting-edge integration concepts. Furthermore, advanced simulation of heat dissipation within the package is imperative.

The Center Core Embedding [11] is developed by AT&S and uses an organic laminate substrate to place monolithic components (e.g., RF chips) within an RF-performant multi-layer printed-circuit board (PCB). As depicted in Figure 14, the monolithic components are first embedded within a core dielectric layer. After this, a stackup of metal and dielectric layers is added on the top and bottom of the core layer.

The monolithic integrated components are connected to the first conductive layers by copper plated micro-vias. Micro-vias have a better conductivity compared to solder bonds or aluminum wire bonds, and introduce smaller inductive and capacitive effects, resulting in lower insertion losses for the high-frequency signals.

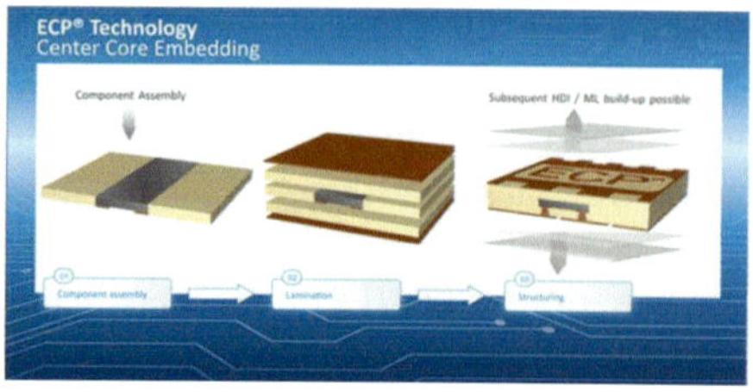

Fig. 14. AT&S Center Core Embedding technology.

The enforcement to integrate chips fabricated using different technologies (BiCMOS and InP) within the same Center Core Embedding presents several challenges. Notably, their thickness must align with the same core layer, and the back-end-of-line (BEOL) finishing of all chips must be compatible with the same micro-via fabrication option. Additionally, the dimensions of the chips are comparable to those of the D-band antenna, posing challenges for floor planning and routing within the RF package itself. These challenges, which are still under study, are pivotal for achieving the functional integrated D-band transceiver envisioned in this project.

F. Antennas in Packaging

The advanced RF packaging technology is essential to ensure a tight and low-loss integration of millimeter-wave integrated circuits (MMICs) and high-efficiency off-chip antennas, as well as an accurate and repeatable manufacturing of the TX and RX module at D-band. The constraints of standard PCB technology, such as the minimum conductor width (~75 μm) and diameter of the vias (~100 μm), and the related tolerances severely limit the design flexibility and the achievable antenna performance beyond 100 GHz. Among the D-band modules reported in the literature, only the one described in [17] uses a standard PCB technology. The MMIC was flip-chipped using solder bumps on the bottom of the PCB and excited a 2 × 2 cavity-backed microstrip patch array on the top side. A moderate sensitivity of the input impedance of the antenna to the assembly process and relatively high interconnect losses (~1.5 dB) were observed. Similar designs with an external MMIC connected to an antenna-in-package (AiP) were realized using High-density interconnect (HDI) PCB technology [18] and semi-additive processing [19], enabling design features down to 40 and 15 μm, respectively. The matching and gain bandwidth of these antennas are generally less than 15% [19]. With respect to these works, the MMICs will be embedded in the core substrate of TX and RX modules, leveraging on the proposed packaging solution, and directly connected to the antennas, reducing the overall occupied area and the interconnect loss. The radiating elements and part of their feeding structures will be manufactured with several stacked pre-preg layers, each with a thickness of 20 μm, enabling the realization of micro-vias with a 20-μm diameter and a minimum conductor width of 15 μm. A main objective is to limit the number of prepreg layers, to enhance the robustness and viability of the modules, achieving at the same time a fractional bandwidth of 10% or greater. To this end, several antenna designs are being developed and compared. Different solutions could be optimized for the TX and RX modules, considering the specific integration and performance requirements

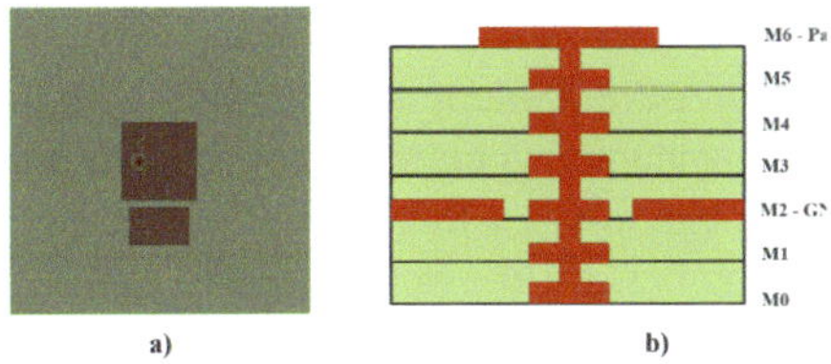

Fig. 15. Probe-fed patch antenna with parasitic: a) top-view of HFSS model, b) cross-section of the antenna stack-up (prepreg layers on top of the core).

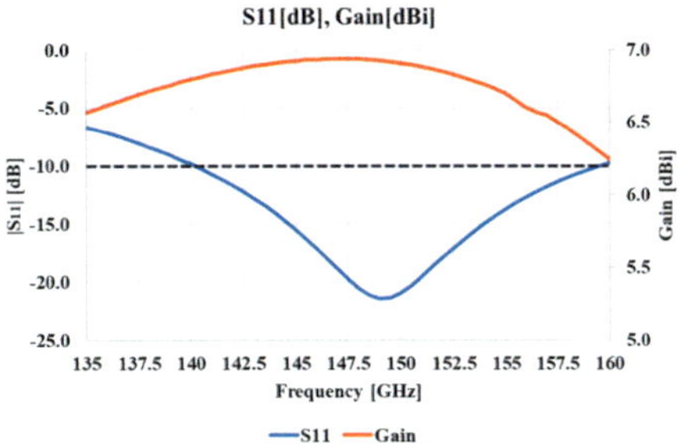

Fig. 16. Simulated S11 and gain of the probe-fed antenna patch with parasitic, in function of frequency.

of each system, as well as the manufacturing capabilities available for the final demonstrators.

Regarding the TX array, a probe-fed patch antenna element has been designed, as part of the initial system-level study. It comprises a single coplanar parasitic radiator to widen the bandwidth. Figure 15a shows the top-view of the simulation model of the design and Figure 15b illustrates the related cross-section, including six 20-μm thick Ajinomoto build-up films, modeled as dielectrics with a relative permittivity of 3.2 and a loss tangent of 0.01. The simulated input reflection coefficient and gain of the stand-alone antenna are plotted in Figure 16 as functions of frequency. The current –10 dB matching bandwidth of the antenna ranges from 140 to 160 GHz with a gain higher than 6 dBi.

In parallel to standard patch antennas, a more complex radiating element, a magnetoelectric (ME) dipole [20], is under development for the RX system. This antenna generally attains a higher gain and a wider bandwidth with respect to patches, for a given electrical distance between the radiator and ground plane. The technological feasibility of a preliminary ME dipole has been validated. The distance between the printed dipole and the ground plane is 120 μm (six pre-preg layers). The ME dipole is fed by a substrate-integrated waveguide in the core substrate. The simulated peak gain is 8.7 dBi at 156 GHz and the –3 dB gain bandwidth spans from 137.5 to 165 GHz.

An additional design iteration is anticipated for both proposed antennas to enhance the performance of the transmitter (TX) and receiver (RX) antenna arrays, as they are influenced by mutual coupling among the elements and finite-size effects. Furthermore, the RF interconnects from the MMIC through the core substrate, require thorough consideration. Particularly critical is the optimization of these transitions, especially for the InP PAs of the transmitter.

Finally, a second design of the antenna-in-package (AiP) is envisioned to co-optimize the impedance of each antenna element with the impedance of the corresponding input (output) of the RX (TX). This co-design may be greatly beneficial, with respect to standard conjugate matching approaches, to enhance the SNR of the RX module and the scanning range and EIRP of the TX array.

V. CONCLUSION AND OUTLOOK

This work focuses on the development of an innovative transceiver prototype suitable of future generation mobile communication. Leveraging the broad spectral bandwidth beyond 100 GHz, this transceiver targets to utilize two 5-GHz RF channels at D-band, namely 143.5–148.5 GHz and 151.5–156.5 GHz. We aim to demonstrate wireless beamformed links between a TX phased-array front-end and a RX phased-array front-end.

An initial study of the link budget is presented, envisioning LoS communication between a BS and a UE, achieving data rates of up to 30 Gb/s over a distance of 100 m. The specifications derived for the upstream transmission from the UE and downstream reception at the UE guide the circuit design process. However, the final demonstrator aims for a less demanding wireless link scenario. This approach is adopted to mitigate risks and allows us to concentrate on addressing the critical technical challenges associated with technology integration.

The RX uses ST BiCMOS B55X technology while the TX combines chips in ST B55X and III-V-Lab InP technologies. The B55X technology boasts top-of-the-line performance for RF and mixed signal applications on silicon, whereas the InP technology excels in efficient amplification and power delivery beyond 100 GHz. Integration of both TX and RX front-ends will be accomplished within an advanced multi-layer RF packaging, featuring antennas that promise low-loss RF performance at D-band. The proposed circuit topologies for the B55X RX and TX front-ends, as well as for the InP PAs, prioritize RF functionality and efficiency. The antennas are designed on one side of the RF advanced packaging stack-up. The TX RF packaging, integrating both B55X and InP chips with distinct BEOL finishing, demands extra attention.

All these preliminary studies and choices guide the continued chip designs and fabrications. Firstly, the tape-out of critical test structures is planned, including test vehicles for the RF packaging. Subsequently, the design of chips and packaging for the final wireless demonstrator will be executed.

ACKNOWLEDGMENT

European project SHIFT is supported by the Chips Joint Undertaking and its members under grant agreement number 101096256.

References

[1] K. K. Tokgoz et al., "A 120Gb/s 16QAM CMOS millimeter-wave wireless transceiver", IEEE International Solid - State Circuits Conference (ISSCC), pp. 168–170, Feb. 2018.

[2] H. Wang, H. Mohammadnezhad, P. Heydari, "Analysis and Design of High-Order QAM Direct-Modulation Transmitter for High-Speed Point-to-Point mm-Wave Wireless Links", IEEE Journal of Solid-State Circuits, vol. 54, no. 11, pp. 3161–3179, nov. 2019

[3] H. Mohammadnezhad, H. Wang, A. Cathelin, P. Heydari, "A 115–135-GHz 8PSK Receiver Using Multi-Phase RF-Correlation-Based Direct-Demodulation Method", IEEE Journal of Solid-State Circuits, vol. 54, no. 9, pp. 2435–2448, Sept. 2019

[4] N. Davy et al, "455/610-GHz fT/fMAX InP DHBT technology with demonstration of a beyond-110-GHz-bandwidth large-swing PAM-4 2:1 AMUX-driver", CSW 2023

[5] A. S. Ahmed et al. "A 140GHz power amplifier with 20.5dBm output power and 20.8% PAE in 250-nm InP HBT technology," IMS 2020.

[6] P. Chevalier et al., "SiGe BiCMOS Current Status and Future Trends in Europe," 2018 IEEE BiCMOS and Compound Semiconductor Integrated Circuits and Technology Symposium (BCICTS), San Diego, CA, USA, 2018, pp. 64–71

[7] C. Desset, N. Collaert, S. Sinha and G. Gramegna, "InP / CMOS co-integration for energy efficient sub-THz communication systems," 2021 IEEE Globecom Workshops (GC Wkshps), Madrid, Spain, 2021, pp. 1–6

[8] B. Sadhu, X. Gu and A. Valdes-Garcia, "The More (Antennas), the Merrier: A Survey of Silicon-Based mm-Wave Phased Arrays Using Multi-IC Scaling," in IEEE Microwave Magazine, vol. 20, no. 12, pp. 32–50, Dec. 2019

[9] [ETSI13810102v150600] https://www.etsi.org/deliver/etsi_ts/138100_138199/13810102/15.06.00_60/ts_13810102v150600p.pdf

[10] C. J. Hansen, "WiGiG: Multi-gigabit wireless communications in the 60 GHz band," in IEEE Wireless Communications, vol. 18, no. 6, pp. 6–7, December 2011, doi: 10.1109/MWC.2011.6108325

[11] EP2342958B1 "METHOD FOR INTEGRATING AN ELECTRONIC COMPONENT INTO A PRINTED CIRCUIT BOARD"

[12] https://docdb.cept.org/download/a5533f97-5a92/Rec1801.pdf

[13] Adler, R., "A Study of Locking Phenomena in Oscillators," Proceedings of the IRE, vol.34, no.6, pp. 351–357, June 1946 E. P. Wigner, "Theory of traveling-wave optical laser," Phys. Rev., vol. 134, pp. A635–A646, Dec. 1965.

[14] N. Ebrahimi, P. -Y. Wu, M. Bagheri and J. F. Buckwalter, "A 71–86-GHz Phased Array Transceiver Using Wideband Injection-Locked Oscillator Phase Shifters," in IEEE Transactions on Microwave Theory and Techniques, vol. 65, no. 2, pp. 346–361, Feb. 2017

[15] B. -W. Min and G. M. Rebeiz, "Single-Ended and Differential Ka-Band BiCMOS Phased Array Front-Ends," in IEEE Journal of Solid-State Circuits, vol. 43, no. 10, pp. 2239–2250, Oct. 2008

[16] F. Qiu, H. Zhu, W. Che and Q. Xue, "A K-Band Full 360° Phase Shifter Using Novel Non-Orthogonal Vector Summing Method," in IEEE Journal of Solid-State Circuits, vol. 58, no. 5, pp. 1299–1309, May 2023

[17] A. Hamani et al., "A D-band multichannel TX system-in-package achieving 84.48 Gb/s with 64-QAM based on

45-nm CMOS and low-Cost PCB technology," IEEE Trans. Microw. Theory Techn., vol. 70, no. 7, pp. 3385–3395, July 2022.

[18] A. Bisognin et al., "Ball grid array module with integrated shaped lens for 5G backhaul/fronthaul communications in F-band," IEEE Trans. Antennas Propag., vol. 65, no. 12, pp. 6380–6394, Dec. 2017.

[19] D. del Rio et al., "A D-band 16-element phased-array transceiver in 55-nm BiCMOS," IEEE Trans. Microw. Theory Techn., vol. 71, no. 2, pp. 854–869, Feb. 2023.

[20] M. Li and K. -M. Luk, "Wideband magneto-electric dipole antenna for 60-GHz millimeter-wave communications," IEEE Trans. Antennas Propag., vol. 63, no. 7, pp. 3276–3279, July 2015.

Giovanni Mangraviti received the B.Sc. and M.Sc. degrees in electrical engineering from the University of Messina (Università degli Studi di Messina), Italy, respectively, in 2005 and 2008, and the Ph.D. degree from the Free University of Brussels (Vrije Universiteit Brussel), Belgium, in 2015.

In 2007, he was an Intern with IMEC, Leuven, Belgium, where he was involved with the characterization and investigation of advanced low-k dielectric materials for back-end-of-line. Since 2009, he has been with the mmWave R&D Team of IMEC. Since 2019, he has been serving as TPC member of the *IEEE Asian Solid-State Circuits Conference*.

Investigation, design, measurement, and advance of mmWave-integrated core circuits and transceivers currently involve his research activities and passion.

Cedric Dehos joined CEA Leti, Grenoble, France, in 2003 after graduation from ESIEE Paris (Dipl.-Ing. and M.Sc.) in telecommunication and signal processing. Since then he has been involved in system level design of complex RF and digital base band circuits including RF behavioral modeling, architecture design and specification, system simulation, and base band signal processing. His main topics of interest included ultra-wide band systems, digital compensation of RF impairments, and antenna beamforming. Since 2006, he has been involved in various mmWave developments in CMOS or BiCMOS technologies, including 24/79GHz short-range radar, 60GHz WiGig. In 2011, he led Leti mmWave designs and began moving the developments towards 5G small cells, promoting the use of mmWave bands for 5G, and towards short-range chip-to-chip communications and contactless connectors. More recently, he proposed an architecture for channel aggregation in D-band, targeting >100 Gb/s wireless communication. He has been involved in many collaborative projects funded by European Commission, as well as in many industrial bilateral projects with transfer of technology. He is also active in the promotion of mmW wireless technologies for particle physics with CERN.

Vincent Puyal received the B.S. degree in electronics engineering from Polytech Montpellier, University of Montpellier, France, in 2002 and the Ph.D. degree in microelectronics engineering from the University of Montpellier, France, in 2007. From 2002 to 2006, he was involved in the design and testing of high-speed bipolar integrated circuits (ICs) for optical communication systems in collaboration with the LIRMM, Montpellier, France and Alcatel-Thales III-V Lab, Marcoussis, France. From 2007 to 2011, he was RF MEMS component and IC research engineer at the CNRS LAAS, University of Toulouse, France. In 2011, he joined the CEA LETI in Grenoble, France. His current research interests include RF and mmW front-end Si ICs for Satcom, 5G and 6G, cryogenic control and readout ICs for quantum computing, EM and signal integrity modeling for 3D sequential integration (3DSI) and high-performance back-end RF and mmW switches (in particular, GeTe and BiCMOS switches).

Olivier Richard received the M.S. degree in microelectronics from the Science University of Orsay, Orsay, France, in 1999, and the Ph.D. degree in microwaves and electromagnetics from the University of Grenoble, Grenoble, France, in 2002. His dissertation focused on the electromagnetic behavioral model of plastic packages for RF applications. He performed his doctoral work at the Institute of Microelectronics, Electromagnetic and Photonics (IMEP), Grenoble, France, and STMicroelectronics, Crolles, France. In 2002, he joined the Technology R&D Department, STMicroelectronics, Crolles, France, as Millimeter Wave and RF Design

Engineer. From 2016, he was successively the project manager, design leader, and team leader in various small companies and startups in the fields of micro-display, RF, and millimeter-integrated circuit design. Since 2023, he has been an expert in RF and mmW design with the French Alternative Energies and Atomic Energy Commission (CEA)-Laboratory of Electronics and Information Technologies (Leti), Research Institute, Grenoble, France.

Ghita Yaakoubi Khbiza holds a Ph.D. in RF and millimeter wave engineering and has two years of experience at CEA-LETI. With over five years in the field, she specializes in designing and optimizing RF and millimeter-wave circuits and systems. Her expertise includes complex simulations, precise layout designs, and thorough chip characterization. Committed to innovation, she drives projects from concept to production, advancing high-frequency technology in collaborative environments.

Francesco Foglia Manzillo (Member, IEEE) received the M.Sc. degree from the University of Naples Federico II, Naples, Italy, in 2012, and the Ph.D. degree from the University of Rennes 1, Rennes, France, in 2017, both in electrical engineering. In 2016, he was a visiting Ph.D. student at Radiation Laboratory, University of Michigan, Ann Arbor, USA. Since July 2017, he has been an antenna scientist with CEA-Leti, Grenoble, France. His research interests include the analysis, synthesis, and design of antenna arrays, quasi-periodic electromagnetic surfaces, numerical modeling, and the integration of millimeter-wave radio systems. Dr. Foglia Manzillo received the Best Applied Technology Paper Award at the 18th European Conference on Antennas and Propagation and the Best Innovation Award at the 39th European Space Agency Antenna Workshop, and co-authored a paper awarded with the Young Engineer Prize at the 51st European Microwave Conference.

Xuan Viet Linh Nguyen received the master's and Ph.D. degrees in electrical engineering from the National Institute of Applied Sciences of Lyon (Institut National des Sciences Appliquées de Lyon – INSA Lyon) in 2019 and 2023, respectively. Since 2023, he has been a Post-Doc Researcher with CEA-Leti, Grenoble, France. His research interests include the analysis and design of rectenna, transmitarray, and reflectarray.

Francesco Filice received the bachelor's degree in electronics engineering and the master's degree in telecommunication engineering from the University of Calabria, Italy, in 2013 and 2016, respectively. From 2017 to 2020, he pursued his Ph.D. in antenna design for mobile satellite telecommunications in STMicrolectronics, Crolles, France and Polytech Nice Sophia, France. From 2020 to 2022, he was a development engineer within the mmWave short-range communication business line in STMicroelectronics, Grenoble, France. Since 2022, he has been a research engineer with Imec, Leuven, Belgium. His research interests include antenna design for satellite applications, mobile handsets, and sub-terahertz applications. A focus of his activity is about electromagnetic and electro-thermal simulations for package and mmWave phased antenna-arrays design.

Ettore Noccetti is a master's student at the University of Pisa, with a bachelor's degree from the same institution. He recently completed an internship at IMEC from March to August 2024, contributing to the SHIFT project. Ettore is an electronics engineer and aims to continue his academic and professional pursuits in this field, focusing on RF electronics and transceiver architectures.

Pierre-Louis Hellier received the M.Eng. degree in integrated electronic systems from Phelma, Grenoble Institute of Technology, Grenoble, France, in 2023. He is currently pursuing the Ph.D. degree with the TIMA Laboratory, Grenoble Alpes University. His research interests include design, modeling, and characterization of millimeter-wave integrated circuit.

Florence Podevin (Member, IEEE) received the grade of engineer in electronics and microelectronics from the École Centrale de Lille, Villeneuve-d'Ascq, France, in 1998, and the M.Sc. degree in electronics and microelectronics and the Ph.D. degree in microelectronics from the Université des Sciences et Technologies de Lille (USTL), Villeneuve d'Ascq, in 1998 and 2001, respectively. In 2001, she joined the Institute of Microelectronics, Electromagnetism and Photonics, Grenoble, France, as an assistant professor. She is currently a professor with the Laboratory "Techniques of Informatics and Microelectronics for integrated systems Architecture," Grenoble, where she has the position of Director Deputee. Her main research interest concerns low-power RF and millimeter-wave architectures with co-design between passive and active devices, performing integrated distributed and lumped passive elements, distributed amplifiers, phase-change material based switches, or co-integration of spintronics devices with microelectronics. She has authored or co-authored more than 130 papers published in international journals and national or international conferences and co-holds five patents.

Sylvain Bourdel received the Ph.D. degree in microelectronics from the National Institute of Applied Science (INSA) of Toulouse in 2000. He was with the LAAS laboratory of Toulouse where he was involved in radiofrequency systems modeling. He was especially focused on spread spectrum techniques applied to 2.45-GHz transceivers. In 2002, he joined the IM2NP in Marseille where he headed the Integrated Circuit Design team. Meanwhile, he was expert for the SCS world-class clusters and the ARCSIS cluster and member of the National Council of Universities (CNU). He joined in 2013 the IMEP-LaHC laboratory and Grenoble-INP as a full professor. From 2018 to 2021, he led the RFIC-Lab laboratory of Grenoble, France. He is now with the TIMA laboratory. He was involved in the steering committees of several CAS conferences and he is actually the secretary of the French CAS chapter. He works on RF and mmW IC design and integration. He particularly focuses on low-cost and low-power applications. His area of research interest also includes design-oriented MOS modeling, UWB, and RFID systems. He is the author and co-author of about 150 referenced IEEE publications.

Andrea Ruffino received the B.Sc. degree (cum laude) in engineering physics from Politecnico di Torino, Turin, Italy, in 2013, the triple joint M.Sc. degree (cum laude) in micro and nanotechnologies for integrated systems from Politecnico di Torino, Institut National Polytechnique de Grenoble (INPG), Grenoble, France, and École Polytechnique Fédérale de Lausanne (EPFL), Lausanne, Switzerland, in 2015, and the Ph.D. degree in microsystems and microelectronics from EPFL in 2021. From 2022 to 2024, he was with IBM Research Zurich, Rüschlikon, Switzerland, working on cryogenic CMOS RFIC transceivers for readout and control of spin and superconducting quantum computers. He is currently an established researcher at ETH Zürich, Zürich, Switzerland working on cryogenic CMOS electronics and RF/microwave/mmWave IC design for communication and sensing. His main research interests include analog/RF/mmWave integrated circuit design, cryogenic CMOS electronics for quantum computing and superconducting single-flux-quantum electronics. Dr. Ruffino is a co-recipient of the Best Joint Paper Award at the European Solid-State Circuits Conference (ESSCIRC) 2022 and is also a recipient of the IEEE Solid-State Circuits Society (SSCS) Predoctoral Achievement Award for 2020–2021.

Kyung-Sik Choi (Member, IEEE) received the B.S., M.S., and Ph.D. degrees in electrical engineering from the Korea Advanced Institute of Science and Technology (KAIST), Daejeon, South Korea, in 2016, 2018, and 2022, respectively. He is currently a post-doctoral researcher with ETH Zurich, Zürich, Switzerland. His research interests include radio frequency (RF) integrated circuits and systems, and high-frequency (terahertz) integrated circuits and systems based on complementary metal–oxide–semiconductor (CMOS) technology.

Basem Abdelaziz Abdelmagid (Student Member, IEEE) received the B.Sc. and M.Sc. degrees from the Electronics and Electrical Communications Engineering Department, Cairo University, Giza, Egypt, in 2018 and 2021, respectively. He is currently working toward the Ph.D. degree at the Department of Information Technology and Electrical Engineering (D-ITET), Swiss Federal Institute of Technology Zürich (ETH Zürich), Zürich, Switzerland. His current research interests include mmWave and sub-THz integrated circuits and systems.

Mohamed Eleraky (Student Member, IEEE) received the B.Sc. and M.Sc. degrees from Ain Shams University, Egypt, in 2016 and 2020, respectively. He also received the M.Sc. degree from the Georgia Institute of Technology in 2022. He is currently pursuing the Ph.D. degree at ETH Zürich, Switzerland. His research interests include THz/millimeter-wave integrated circuits and systems.

Hua Wang (Fellow, IEEE) is a Full Professor and the Chair of Electronics with the Department of Information Technology and Electrical Engineering (D-ITET) of Swiss Federal Institute of Technology Zürich (ETH Zürich). He is the Institute Deputy Head of the Integrated Systems Laboratory (IIS) at ETH Zürich. He is the Director of the ETH Integrated Devices, Electronics, and Systems (IDEAS) Group. He is a faculty member of the ETH Zürich Quantum Center. Prior to that, he was a tenured associate professor with the School of Electrical and Computer Engineering (ECE) at Georgia Institute of Technology, USA. He held the Demetrius T. Paris professorship at Georgia Tech. He was the director of the Georgia Tech Electronics and Micro-System (GEMS) lab. He worked at Intel Corporation and Skyworks Solutions from 2010 to 2011. He received the M.S. and Ph.D. degrees in electrical engineering from the California Institute of Technology, Pasadena, in 2007 and 2009, respectively. Dr. Wang is interested in innovating analog, mixed-signal, RF, and mmWave integrated circuits and hybrid systems for wireless communication, sensing, and bioelectronics applications. He has authored or co-authored over 250 peer-reviewed journals and conference papers. Dr. Wang is a Top Contributing Author to the IEEE International Solid-State Circuits Conference (ISSCC) for the past 70 years (1954–2023). He received the DARPA Director's Fellowship Award in 2020 (the first awardee in Georgia Tech's history), the DARPA Young Faculty Award in 2018, the National Science Foundation CAREER Award in 2015, the Qualcomm Faculty Award in 2020 and 2021, the IEEE MTT-S Outstanding Young Engineer Award in 2017, the Georgia Tech Sigma Xi Young Faculty Award in 2016, the Georgia Tech ECE Outstanding Junior Faculty Member Award in 2015, and the Lockheed Dean's Excellence in Teaching Award in 2015. Dr. Wang was a Technical Program Committee (TPC) Member for IEEE ISSCC, RFIC, CICC, and BCICTS conferences. He was a Steering Committee Member for IEEE RFIC and CICC. He was the Conference Chair for CICC 2019 and Conference General Chair for CICC 2020. He is a Distinguished Microwave Lecturer (DML) for the IEEE Microwave Theory and Techniques Society (MTT-S) for the term of 2022–2024. He was a Distinguished Lecturer (DL) for the IEEE Solid-State Circuits Society (SSCS) for the term of 2018–2019. He served as the Chair of the Atlanta's IEEE CAS/SSCS joint chapter that won the IEEE SSCS Outstanding Chapter Award in 2014.

Bertrand Ardouin received the M.S. and Ph.D. degrees in electrical engineering from the University of Bordeaux, Bordeaux, France, in 1998 and 2001, respectively. In 2002, he was the Co-Founder of XMOD Technologies, Bordeaux, a startup he has led as a CEO and an R&D Director until 2019. He joined SERMA Technologies, Pessac, France, where he was a Business Unit Manager in charge of electrical expertise, environmental tests, and compact modeling. He joined Nokia Bell Laboratories and III-V Lab in 2023, where he is currently responsible of the High-Speed Analogue Digital Interfaces Group, in Palaiseau, France. Dr. Ardouin's areas of interest cover CMOS, SiGe, and InP HBT devices, harsh-environment electronics, reliability, RF and mmWave characterization, modeling & PDK, statistical modeling, process technology, and high-speed IC design.

Tom K. Johansen (Member, IEEE) received the M.S. and Ph.D. degrees in electrical engineering from the Technical University of Denmark, Kongens Lyngby, Denmark, in 1999 and 2003, respectively. In 1999, he joined the Electromagnetic Systems Group, DTU Elektro, Technical University of Denmark, where he is currently an associate professor. From September 2001 to March 2002, he was a visiting scholar with the Center for Wireless Communication, University of San Diego, San Diego, CA, USA. He has spent several external research stays with the Ferdinand-Braun-Institut (FBH), Berlin, Germany. His current research interests include the modeling of high-frequency solid-state devices, and microwave, millimeter wave, and submillimeter-wave integrated circuit design. Dr. Johansen is a member of the IEEE.

Luca Fanori was born in Pavia, Italy, in 1984. He received the M.Sc. degree in electronic engineering and the Ph.D. in electrical engineering from the University of Pavia, Pavia, Italy. He was in an internship at Marvell Italy and Ericsson AB in Sweden to design All-Digital-PLL and studying LTE transceiver architectures. After two years of post doc at Lund University, he currently works for Marvell Italy, where his research activities are focused on the implementations of frequency synthesizers. Dr. Fanori was a co-recipient of the Best Invited Paper Award at the IEEE 2011 Custom Integrated Circuits Conference (CICC).

Qiuting Huang (Fellow, IEEE) received the Ph.D. degree in applied sciences from the Katholieke Universiteit Leuven, Leuven, Belgium, in 1987. Between 1987 and 1992, he was a lecturer with the University of East Anglia, Norwich, U.K. Since January 1993, he has been with Integrated Systems Laboratory, Swiss Federal Institute of Technology (ETH), Zurich, Switzerland, where he is a professor of Electronics. In 2007, he was also appointed as a part-time Cheung Kong Seminar Professor by the Chinese Ministry of Education and the Cheung Kong Foundation and has been affiliated with the South East University, Nanjing, China. His research interests span RF, analog, mixed analog–digital, as well as digital application-specific integrated circuits and systems, with an emphasis on wireless communications applications in recent years. Dr. Huang is a member of the technical program committee of the European Solid-State Circuits Conference (ESSCIRC). He also served on the technical program and executive committees of the International Solid-State Circuits Conference (ISSCC) between 2000 and 2010.

E. Schlaffer is an Austrian citizen with a technical college for electrical engineering in Kapfenberg. He joined as Process Engineer of AT&S in December 1998 and was responsible as Department Head for mechanical drilling, laser drilling, routing, and electrical test. From 2008 to 2012, he was responsible for the development of a novel Photovoltaic Module Programme. Since 2012, he has been developing concepts for high-speed and radio frequency applications as a program manager in R&D.

Piet Wambacq received the M.Sc. degree in electrical engineering and the Ph.D. degree from Katholieke Universiteit Leuven, Leuven, Belgium, in 1986 and 1996, respectively. In 1996, he joined imec, Leuven, where he is currently Fellow, working on analog/RF/mmWave IC design in various technologies for wireless applications. Since 2000, he has also been a part-time professor with Vrije Universiteit Brussel (VUB), Brussels, Belgium. He has authored or co-authored six books and more than 350 articles in edited books, international journals, and conferences. He has been a member of the Program Committee of ISSCC from 2012 to 2020 with the role of RF Subcommittee Chair from 2016 to 2020. He has been chair of the program committee of ISSCC 2023.

Dominique Morche received the Master of Science (M.S.) degree in engineering from the Ecole Nationale Supérieure d'Electricité et de Radioelectricité de Bordeaux, Bordeaux, France, in 1990 and the Ph.D. degree in electronics from the Institut National Polytechnique de Grenoble, France, in 1994. His Ph.D. mainly focuses on sigma–delta ADCs. From 1994 to 2001, he was with France Telecom, Meylan, France, as a research engineer. He has been involved in the architecture and design of analog circuits for telecom applications. He is currently with Commissariat à l'énergie atomique-Leti Minatec, Grenoble, France where he is a research director. His current field of research is in the specification and design of RF architecture for UWB, mmWave, and IoT systems.

Björn Debaillie leads imec's collaborative R&D program on cutting-edge IoT technologies, covering high-speed communications, high-resolution sensing, and neuromorphic computing. As a seasoned researcher and manager, he is responsible for strategic collaborations and partnerships, innovation management, and public funding policies as well as the operational management and coordination across imec's collaborative programs and projects. Björn Debaillie coordinates public funded projects and seeds new initiatives. He holds patents, has received awards, and has authored books and international papers published in various journals and conference proceedings.

Ka-Band GaN-on-SiC Power Amplifier for High EIRP Satellite Phased Antenna Array

Francesco Manni, Paolo Colantonio, Rocco Giofrè, Ernesto Limiti, Patrick Ettore Longhi, Steven Caicedo Mejillones, Stefano Moscato, and Alessandro Fonte

Abstract—**This work presents a transmitter-only phased antenna array architecture for Ka-band satellite applications, specifically from 25.5 to 27 GHz. The design targets an optimum design point among the number of elements, the output power from each transmitter, the antenna EIRP, and its steering capabilities. The core of the system relies on a custom designed power amplifier based on AlGaN/GaN epitaxy grown on 4-inch SiC substrate featuring 20 dB of gain and 28 dBm of saturated power. The antenna array front-end is envisaged as a lattice of 64 hollow metallic horns, spaced by 16 mm from each other. The amplifiers paired with the antenna array targets 70 dBm of EIRP, even considering the maximum beamsteering configuration of ±20°. The proposed concept highlights the flexibility in targeting different power levels, antenna gain, and steering capabilities to finally address a large number of use-cases.**

This work was supported by the Chips Joint Undertaking (Chips JU) under grant agreement number 101096256 (project SHIFT). The Chips JU receives support from the European Union's Horizon Europe research and innovation program and the National Authorities.

Francesco Manni, Paolo Colantonio, Rocco Giofrè, Patrick Ettore Longhi, and Ernesto Limiti are with the Electronic Engineering Department at the University of Roma "Tor Vergata," Rome, Italy (e-mail: longhi@ing.uniroma2.it). Steven Caicedo Mejillones, Stefano Moscato, and Alessandro Fonte are with the R&D Department, SIAE MICROELETTRONICA, Cologno Monzese (MI), Italy (e-mail: stefano.moscato@siaemic.com).

Index Terms—**Antenna array, Ka-band, phased array, power amplifier, mmWave**

I. INTRODUCTION

AMONG the characteristics of a transmitting phased antenna array (PAA), the effective isotropic radiated power (EIRP) is one of the most important. In fact, the received signal-to-noise ratio is mostly affected by the EIRP at the transmitter side of the link. However, the maximization of both the amount of radiated power and the antenna gain implies trade-offs, which can limit the applicability of the array system itself. For example, an enhancement of the antenna gain leads to a cumbersome reflector or a large array footprint, which can be an issue for both terrestrial and flying transceivers.

The millimeter wave (mmWave) front-end for transmitters on-board to satellites and spacecrafts is generally demanded to a travelling wave tube amplifier paired with a low gain antenna element. The aforementioned amplification stage has excellent performance in terms of power handling, linearity, efficiency, and reliability, proven since decades. However, motors and actuators are generally exploited to allow a fine compensation of the beam direction with disadvantages in terms of on-board mechanical hindrances, weight, reliability, and costs.

Phased antenna array (PAA) is proven to be the solution since it is fully customizable [1–4], scalable in EIRP and frequency, and can be even flat, overcoming the adoption of standard parabolas or isoflux antennas. Basic elements can be fed or clustered, balancing EIRP and steering capabilities [5].

However, PAAs exhibit a main drawback in use-cases where high EIRP is required and the available power supply is limited. The adoption of a large number of radio frequency (RF)-integrated circuits leads to a significant amount of power drawn from the system. The main responsible is clearly the final amplification stage where the power added efficiency (PAE) is the main feature to be taken into account in PAA systems. Not less important, PAE tends to drop for higher frequencies, but in modern systems, the trend is to push the connectivity in the mmWave range to exploit wider channels and, in turn, higher data throughput at a given constellation scheme [6–8].

Power amplification stages based on GaN on SiC have been proved to be the perfect candidate when the power efficiency is paramount, even above the standard microwave frequency bands [9]. The envisaged architecture of PAA perfectly suits the adoption of power amplification stages based on GaN since it opens the possibility to keep high the output power level without an exponential growth of the overall power consumption of the system.

The concept introduced with this manuscript is named ESKaA, which stands for "Electronically Steerable Ka-band Antenna." ESKaA exploits a custom-designed GaN-on-SiC PA and a uniform lattice of 64 horns as radiating elements where the main features are reported in Table I. This manuscript is organized in five sections: Section II introduces the most important trade-off in the phased array design, Section III describes the design of the PA, and Sections IV and V show the system architecture and the antenna performance, respectively. Section VI concludes the paper.

TABLE 1.
Specifications Based on the Identified Use-Case.

Specification	
Operational net bandwidth	25.5–27 GHz
Polarization	Circular
EIRP (guaranteed at max scan angle)	>70 dBm
Scan angle (free from grating lobes)	±20°
DC power consumption	<120 W
Dimensions (max footprint)	200 × 200 mm^2

II. TRADE-OFF IN THE PHASED ANTENNA ARRAY DESIGN

This section describes an effective way to preliminarily estimate the array order N and the recommended output power P_{TX} on each amplifier. The method considers, as input, the required EIRP and the gain of the basic radiating element. To do so, the starting point is the definition of EIRP in dBm, which can be calculated by the following equation:

$$EIRP = 10\log_{10}(N) + G + 10\log_{10}(N) + P_{TX}$$

where G is the gain of the base-block element in decibel and P_{TX} is expressed in dBm. Here it is assumed that each base-block antenna is fed by a single active component. Another assumption is that the base antenna is a circularly polarized horn with 12 dBi of gain. Considering the above-mentioned assumptions, Figure 1 shows a trade-off between P_{TX}, N, and EIRP. The more EIRP is needed, the greater the array order or output power

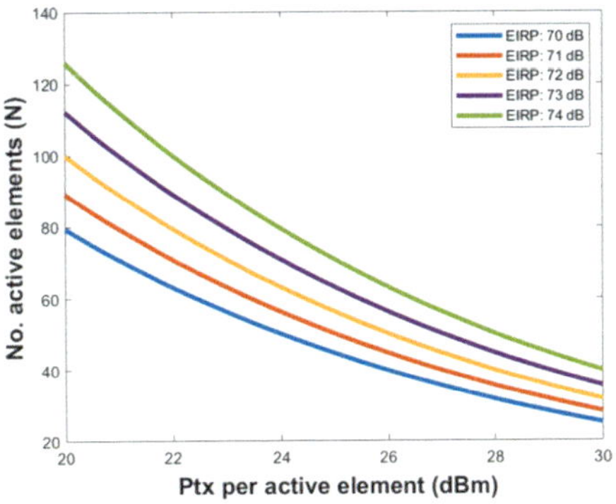

Fig. 1. EIRP, N, P_{TX} trade-offs in the considered phased array antenna architecture.

TABLE 2.
TRADE-OFF IN TARGET EIRP ($G = 12$).

EIRP (dBm) =	**70**	**71**	**72**	**73**	**74**
P_{TX} for N = 7 × 7	24	25	26	27	28
P_{TX} for N = 8 × 8	22	23	24	25	26
P_{TX} for N = 9 × 9	20	21	22	23	24
P_{TX} for N = 10 × 10	18	19	20	21	22

required, as expected. Considering a square array (N), the data extrapolation from the graphs in Figure 1 can be carried out and it is reported in Table II.

The minimum 70 dBm of EIRP must be guaranteed even when the beam is steered at ±20°. Since a maximum scan loss is expected in the order of 2 dB of steering, the minimum EIRP must be updated to 72 dBm.

In order to have some margin, the combination with EIRP = 74 dBm, with $N = 8 \times 8$ elements and P_{TX} = 26 dBm is chosen for the PAA design. The choice was also motivated considering the achievable average output power P_{TX} of the designed GaN-on-SiC power amplifier (with the hypothesis of 2.5 dB of back-off for a 64-APSK modulation scheme). More details of the amplifier are given in the following section. This preliminary analysis serves to initially size the complexity of the design. The final parameters may change depending on the challenges encountered during the actual design of the amplification stages, as well as the antenna system.

III. GAN-ON-SIC POWER AMPLIFIER

The GaN on SiC power amplifier is designed using UMS foundry's GH15 process, based on AlGaN/GaN epitaxy grown on 4-inch SiC substrate. The transistor features a T gate of 150 nm length, whose shape and fabrication processes have been optimized for ensuring good reliability properties. A source-terminated field-plate is added to reduce the feedback capacitance for a higher gain at RF frequencies. The technology shows a DC saturated drain current around 1.2 A/mm and a DC maximum transconductance (g_m) close to 400 mS/mm.

The substrate is thinned to 70 μm by means of a back-side process that ensures the compatibility with both soldering and gluing-based assembly processes. Other figures of merit of the technology are power density at 30 GHz around 3.5 W/mm and a maximum oscillation frequency (f_{max}) above 80 GHz. These figures demonstrate the technology is well-suited for Ka-band applications. More process details are found in [10] and [11].

The PA design targets, in the 25.5–27.0 GHz band, are 20 dB gain, an output power higher than +25.5 dBm and 30% PAE, to be maximized within the technology limitations.

The first step in PA design is to size the total periphery of the final stage transistor. This step is performed considering the required output power (>25.5 dBm) and considering approximately 1 dB resistive loss of the output matching network (OMN). This leads to approximately 27 dBm at device level (i.e., 500–600 mW). The selected technology can provide a power density of 3.5 W/mm

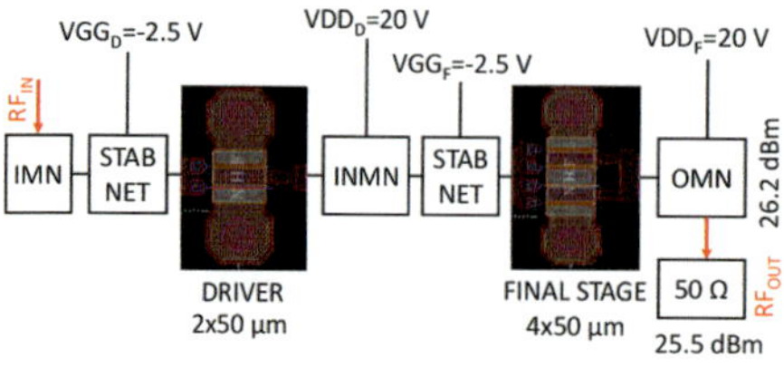

Fig. 2. Simplified PA block diagram.

at +25 V drain voltage. However, we selected to operate the device at a more conservative 3.0 W/mm and +20 V drain voltage to apply space derating conditions, therefore improving reliability. As a result, the total periphery of the final stage is 0.2 mm. A four-finger by 50 µm device is selected to balance the competing requirements of output power, efficiency, and gain. The gain of the final stage is computed to be 10 dB and a two-stage topology is required to achieve a gain greater than 20 dB, sketched in Figure 2.

Another reason for lowering the drain voltage to +20 V is to improve efficiency. Furthermore, the PAE of the single device is approximately 40%; so realistically it is only possible to achieve 30% of PAE in a two-stage PA.

The OMN (Figure 2) is designed to exhibit a purely resistive load at the intrinsic voltage-controlled current source of the final stage transistor. Likewise, to improve efficiency, short circuits are imposed on the higher order harmonics at the same section of the circuit.

The size of the driver (first stage) Field Effect Transistor (FET) is half that of the final stage, as a compromise of linearity, gain, and efficiency at amplifier level. The small-signal gain of active devices is approximately 17 dB while the cumulative loss of passive structures is 3 dB.

Stability networks (STAB NET in Figure 2) are inserted in the HPA to guarantee device unconditional stability over a very wide range of frequencies, practically from 100 MHz to f_{max}.

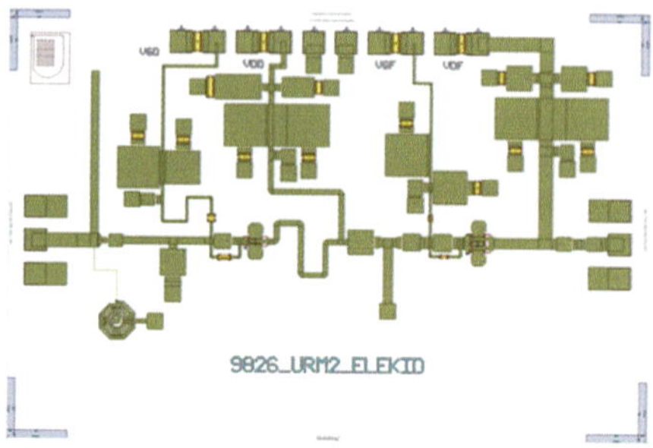

Fig. 3. PA layout. Chip size is 3.0 mm × 2.0 mm.

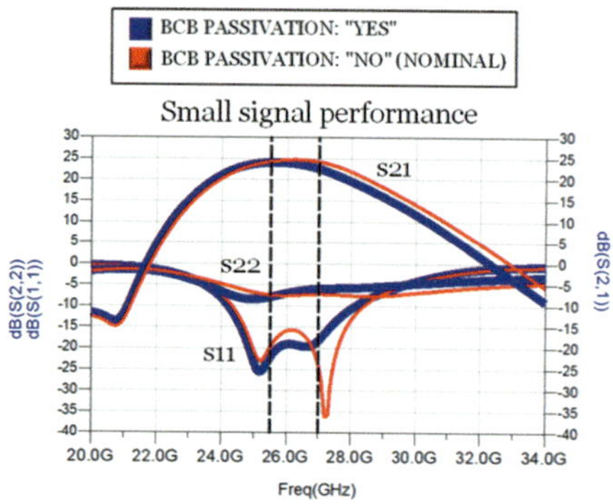

Fig. 4. PA simulated S-parameters. The dashed vertical lines define the operating bandwidth.

Finally, the input matching network and Inter-stage Matching Network (IMN) are designed to provide gain/power terminations at the considered active device's terminals in the operating bandwidth.

The PA's layout is provided in Figure 3. The size is 6 mm^2 and is essentially fixed by the foundry's multi-project wafer run constraints. A smaller chip footprint could have been obtained in a dedicated wafer run. The simulated PA linear performance is shown in Figure 4. The blue traces represent simulations using the benzocyclonbute (BCB) option, while the red traces indicate when the nominal (qualified) manufacturing process is applied. For this manufacturing iteration, we opted for BCB protection, to improve chip ruggedness in harsh operating environments. The DC gate voltage is adjusted around

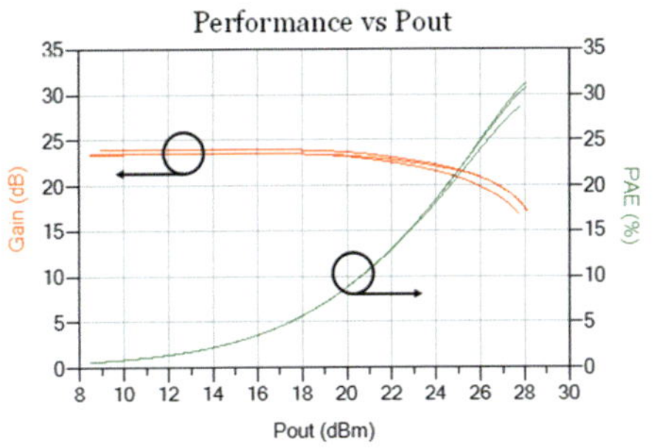

Fig. 5. PA simulated gain and PAE as a function of output power. Three traces are provided for each parameter at 25.5, 26.25, and 27.0 GHz.

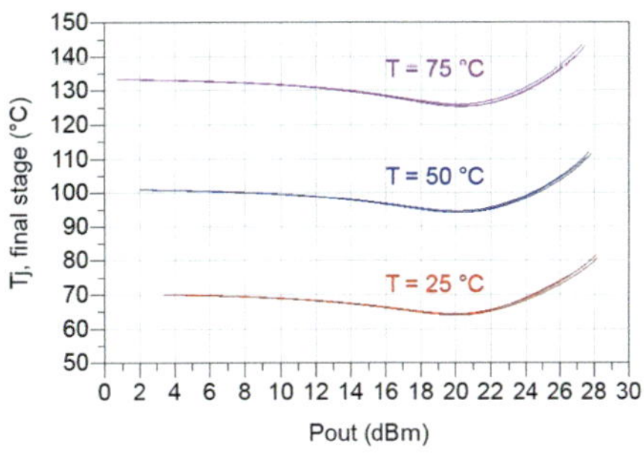

Fig. 6. PA final stage HEMT junction temperature as a function of output power and base-plate temperature 25/50/75 °C. Three traces are provided for each set of data at 25.5, 26.25, and 27.0 GHz.

−2.4 V to obtain an overall DC drain current, in linear operating mode, of 60 mA (20 mA on the driver and 40 mA on the final stage). The DC drain voltage is set at +20 V as per design choice.

The gain is well above 20 dB in the range of 25.5–27.0 GHz. As shown in Figure 4, and the input matching is better than 15 dB, while the output matching is between 7 and 10 dB depending on the selected production process.

The PA's simulated nonlinear performance is provided in Figure 5. The saturated output power reaches +28 dBm, therefore fulfilling the >25.5 dBm requirement. For this high output power level, the PAE reaches 30% while requiring more than +10 dBm input power. If the PA is operated in back-off, around +25.5 dBm output power, then the PAE is reduced to 23%. The advantage is that a significantly lower RF input power is required, which is only +5 dBm. The overall DC drain current reaches 80 mA at +26 dBm output power and 100 mA at saturated output power (+28 dBm).

Lastly, the final stage HEMT junction temperature (T_j) is plotted in Figure 6. This parameter is critical to ensure adequate device reliability. Recent indications by the European Space Agency (ESA) reported in [12] recommend this value should remain below +160 °C to ensure adequate reliability in components employed in spaceborne payloads. This channel temperature operation is obtained by design, applying the synthesis methodology reported in [13] where the classical power amplifier design paradigm is reconsidered shifting to a thermal-aware design approach. The MMIC has completed the manufacturing process and is currently being tested to verify the compliance with design targets and simulated data.

IV. ESKAA SYSTEM ARCHITECTURE

The proposed concept is envisaged as a 64-element transmitting only PAA operating in the lower portion of the Ka-band. Table I summarizes the most important features of the intended antenna system, but this section deals with the foreseen architecture of the overall printed circuit board (PCB) and antenna assemblies, sketched as a simplified block diagram in Figure 7.

The phased array is based on a complex electronic system designed onto a 12-metal layer 200 mm × 200 mm PCB. The copper clads and layers are mixed per type and thickness to accomplish different tasks. The top layer is

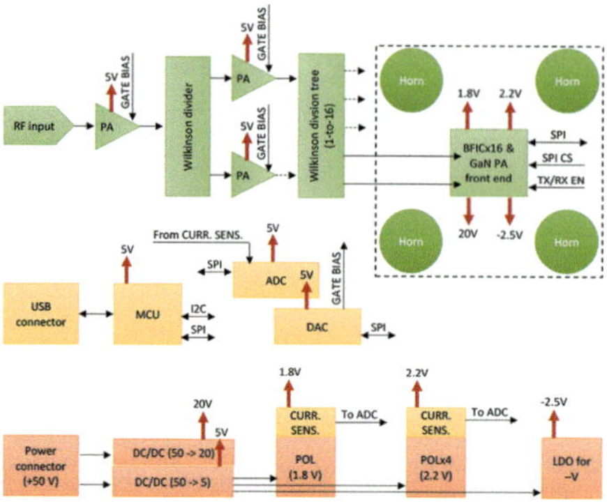

Fig. 7. System block diagram of the phased array. RF paths (green), digital and monitoring functionalities (yellow), and power tree (red) have been highlighted in this sketch.

principally devoted to route the Ka-band signal. A 4-mil Rogers 4350B LoPro has been chosen as premium RF material with a 17 μm copper foil featuring a very low surface roughness. Down to the fifth copper layer, the chosen dielectrics are larger in thickness but very similar to the top one per electrical prorieties. This choice guarantees optimal performance in the microstrip-to-waveguide launcher, exploited as feeding system for each horn antenna. From fifth to the bottom layer, the dielectrics are FR4-like since no RF signal is routed. The copper clads of inner layers have been chosen accordingly to their main functionalities: layers 3 and 6 are 35 μm thick to route the +5 V supply up to 20 A, layers 9 and 11 are 70 μm thick to supply the beamformer ICs with the +1.8 V and +2.2 V high-current rails, and the remaining layers are 17 μm thick to mainly support digital signals and medium-to-low power supplies.

The RF architecture relies on single feeding point, designed as a 2.92 mm end-launch connector. Then, a planar splitter tree based on a cascade of custom-designed Wilkinson divider and the series of two gain block feeds with 16 ICs. Each chip has two common input ports and is foreseen to be connected to the final PA stage and serves a subarray of 2 × 2 horns. The overall network embeds 31 splitters loaded with 0201 commercial 100 Ohm resistors, rated up to 50 GHz. Given the top-layer physical and electrical features, the calculated loss per unit length of the 50 Ohm microstrip has been computed as 24 dB/m with an overall routing of 200 mm. The analog front-end finally exploits the GaN-based PA, described in the previous section, which directly feeds the antenna through ad-hoc launchers.

The PAA design also includes digital features. A microcontroller unit (MCU) is exploited to manage Serial Peripheral Interface (SPI) connection with the beamformer ICs. Digitally controlled multi-channel digital-to-analog convertors are also embedded to finely adjust the analog PA gate biasing and to partition each drain supply for a programmable start-up sequence. Analog-to-digital convertors are used to read current sensors placed along the main supply rails. Standard USB connectors enable the MCU programming through an external device.

A single power supply of +50 V is chosen to fully bias the system. A three-polar connector is mounted directly onto the PCB and connected to the main DC/DC isolated power converters through a capacitor filtering scheme, a fuse in case of short circuit, and a circuit to prevent reverse polarization biasing. The down-conversion drops the supply to +5 and to +20 V. The main +5 V rail is then routed to five non-isolated DC/DC point of load converters. One is devoted to down-convert the +5 to +1.8 V up to 12 A, whereas the other four convert the main rail to +2.2 V up to 8 A each. The main supplies +5 and +20 V directly bias the drains of the RF gain blocks and the final PA stages, respectively. An Low-Dropout regulator

(LDO) is also present to generate the negative supply voltage of −2.5 V, needed for the gate biasing of the power amplifiers.

The arrangement of the components on the circuit board needs to be properly taken into account. The split of PCB layout into different functional areas (RF, digital, and power) is one golden rule in electronic design, especially in a complex system like a PAA. Additionally, the overall assembly needs to accommodate also waveguide structures to feed the matrix of 64 circular horn antennas.

Each microstrip-to-circular waveguide launcher is designed on the top layer of the PCB where the lowest RF losses are expected and where the whole RF circuit is foreseen. However, the launcher needs a circular backshort; then the stack-up thickness has been designed to match exactly the quarter-wavelength dimension needed by the transition. To diminish the losses, lower FR4-like layers are milled in correspondence of each waveguide feed. Contrarily, the upper low-loss dielectric layers fully cover the circular section allowing the penetration of the microstrip launchers inside the waveguide footprint and reinforce the membrane beneath the antenna structure.

The whole PCB is supported by a flat metallic carrier, made by aluminum. A silver finishing is foreseen since each waveguide backshort is sealed by the metallic surface. The carrier is then machined to allow a perfect matching with the components placed on the bottom side of the stack-up. The overall assembly is then closed by a cover, fully machined from a single aluminum block. The cover has multiple functions: it acts as antenna since all the 64 circular horns are engraved into this top part, but it also serves as heatsink for the beamformer ICs and the main DC/DC converter. The overall weight of the bottom and top mechanical parts is in the order of 1.2 kg.

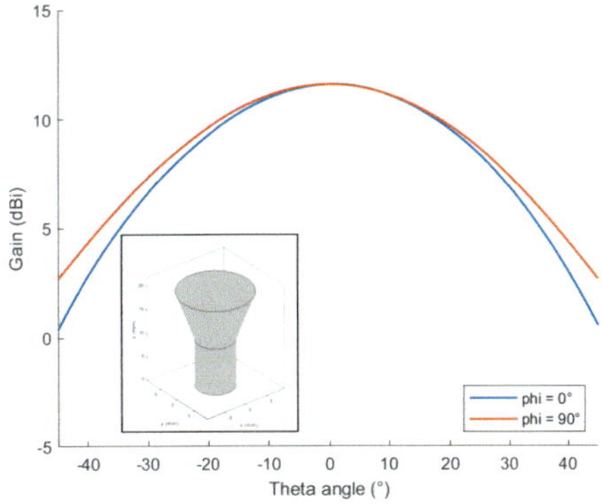

Fig. 8. Directivity cuts for $\Phi = 0°/90°$ and 3D view of the basic radiating element.

V. UNIFORM GRID ANTENNA SIMULATIONS

The first implementation of ESKaA is envisioned as a uniform grid of 8 × 8 circular horns.

The design of the basic antenna has been addressed considering the trade-off in the phased array design. A larger antenna element will result in a larger pitch between each radiator than a higher gain but lower steering capabilities. On the contrary, a matrix of smaller horns can address wider steering angles at the cost of an overall lower array factor gain. The balance has been found considering the regular lattice of 64 horns delivering 12.2 dBi of gain.

The preliminary antenna design follows general rules of circular horn synthesis, well described in [14]. The input waveguide is 8 mm in diameter, whereas the aperture diameter is 15.2 mm. The horn's flare is set as 10 mm, an optimal trade-off between aperture efficiency, input matching and thickness of the foreseen metallic cover, which has to accommodate the antennas. Figure 8 shows the directivity of the basic element estimated through an antenna synthesis. The slight differences in the Half Power Beamwidth (HPBW) between the 0° and 90° cuts are aligned

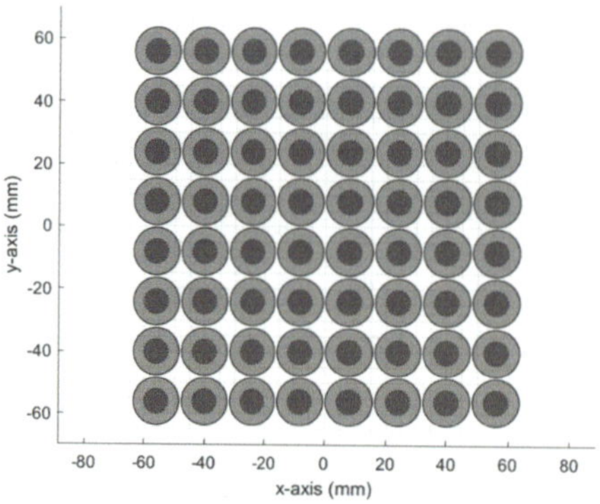

Fig. 9. Lattice of 64-horn antenna elements. Dark gray dotted circles show the footprints of the input circular waveguides, whereas the larger gray circles mimic the 15.2 mm horn apertures.

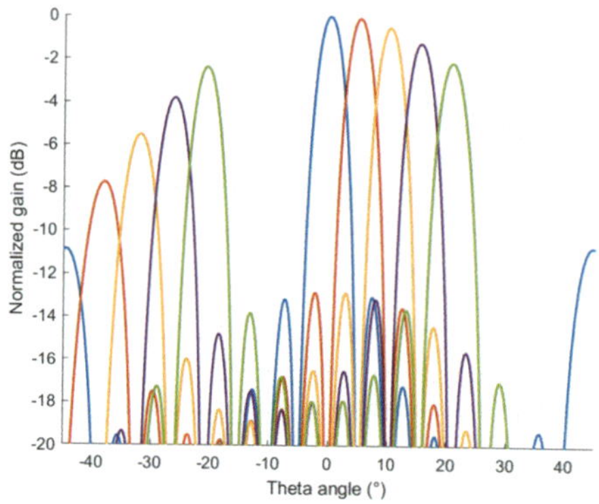

Fig. 10. Array directivity versus theta angle for $\Phi = 0°$ in different steering configurations, from boresight direction to 20° tilt.

with the fact that the formulas refer to a linear polarized feeding. A preliminary 3D view of the antenna is also shown in Figure 8 to highlight the involved aspect ratio between input waveguide diameter, the horn's aperture, and the flare length.

The array is the composition of the basic radiators in a uniform square lattice of 8 × 8 elements, as sketched in Figure 9. The normalized radiation pattern of the array is shown in Figure 10. The maximum gain is however lower with respect to the target since it stops to 26.9 dBi. The overall EIRP of the system is then 71 dBm when combined with the total output power of 26 dBm from the 64 antennas. The graphs in Figure 10 also show the theoretical beamsteering performance of the PAA. The evaluated tilted angle is up to 20°, which is the maximum allowed to be free from grating lobes in the observation range of ±20°. Thanks to the normalization, the scan loss is highlighted and evaluated in 0.11 dB/°. First secondary lobes appear to be −13 dB from the peak gain, without applying any amplitude tapering.

The center-to-center distance of 16 mm, equal to $0.71 \cdot \lambda_0$ calculated at 26.25 GHz, makes the square footprint of the array of 128 mm^2. The compactness of the array perfectly fits the designed square PCB of 200 mm^2. The remaining areas surrounding the antenna-only footprint are devoted to accommodate the power supply elements and the digital architecture.

Preliminary antenna synthesis did not take into account any full-wave simulations. However, the foreseen real implementation of the horn antenna array will be based on CNC machining to guarantee extremely tight accuracy and smooth surface finishing. The horn's flare will be approximated with a 1 mm stepped profile in place of an ideal smooth one. This method allows to significantly reduce the manufacturing complexity and speed up the machining time. Silver plating is also foreseen on top of aluminum to lower ohmic losses involved in the waveguide feeding system and into each hollow antenna.

VI. CONCLUSION

This paper introduces the ESKaA concept, a transmitting only PAA for Ka-band applications. The manuscript gives details in terms of system sizing, envisaged PCBA architecture, and overall performance. Core of the PAA is the GaN PA, specifically designed to match the requirements in terms of output power and PAE, expected to be as high as 28 dBm and 30%, respectively. Simulations of the 64-element antenna array show a maximum boresight gain of 26.9 dB and an addressable steering angle up to 20°.

ACKNOWLEDGMENT

The authors would like to thank the University of L'Aquila, the University of Calabria, the Politecnico of Milano, and MTR S.r.l. for being all together involved in this activity and in charge of the design of the RF building blocks, antenna launchers, antenna array, and the mechanical assembly, respectively.

REFERENCES

[1] X. Luo, et al., "A Scalable Ka-Band 1024-Element Transmit Dual-Circularly-Polarized Planar Phased Array for SATCOM Application," in IEEE Access, vol. 8, pp. 156084–156095, 2020.

[2] K. K. W. Low, S. Zihir, T. Kanar and G. M. Rebeiz, "A 27–31-GHz 1024-Element Ka-Band SATCOM Phased-Array Transmitter With 49.5-dBW Peak EIRP, 1-dB AR, and ±70° Beam Scanning," in IEEE Transactions on Microwave Theory and Techniques, vol. 70, no. 3, pp. 1757–1768, March 2022.

[3] K. K. W. Low, A. Nafe, S. Zihir, T. Kanar, and G. M. Rebeiz, "A scalable circularly-polarized 256-element Ka-band phased-array SATCOM transmitter with ±60° beam scanning and 34.5 dBW EIRP," in IEEE MTT-S Int. Microw. Symp. Dig., Jun. 2019, pp. 1064–1067.

[4] G. Gultepe, T. Kanar, S. Zihir, and G. M. Rebeiz, "A 1024-element Ku-band SATCOM phased-array transmitter with 45-dBW singlepolarization EIRP," IEEE Trans. Microw. Theory Techn., vol. 69, no. 9,pp. 4157–4168, Sep. 2021.

[5] S. C. Mejillones, M. Oldoni, S. Moscato, A. Fonte and M. D'Amico, "Power Consumption and Radiation Trade-offs in Phased Arrays for 5G Wireless Transport," 2020 43rd International Conference on Telecommunications and Signal Processing (TSP), Milan, Italy, 2020.

[6] "5 trends in satellite communications on the horizon", ITU website, Accessed: Oct. 2023, March 2020, https://www.itu.int/hub/2020/03/5-trends-in-satellite-communications-on-the-horizon/.

[7] M. Abo-Zeed, J. B. Din, I. Shayea, and M. Ergen, "Survey on land mobile satellite system: Challenges and future research trends," IEEE Access, vol. 7, pp. 137291–137304, 2019.

[8] S. Burleigh, T. Cola, S. Morosi, S. Jayousi, E. Cianca, and C. Fuchs, "From connectivity to advanced internet services: A comprehensive review of small satellites communications and networks," Wireless Commun. Mobile Comput., vol. 2019, pp. 1–17, May 2019.

[9] R. S. Pengelly, S. M. Wood, J. W. Milligan, S. T. Sheppard and W. L. Pribble, "A Review of GaN on SiC High Electron-Mobility Power Transistors and MMICs," in IEEE Transactions on Microwave Theory and Techniques, vol. 60, no. 6, pp. 1764–1783, June 2012.

[10] PROCESS Corporation, Boston, MA, USA. Intranets: Internet technologies deployed behind the firewall for corporate productivity. Presented at INET96 Annual Meeting. [Online]. Available: http://home.process.com/Intranets/wp2.htp

[11] V. Di Giacomo-Brunel et al., "Industrial 0.15-µm AlGaN/GaN on SiC Technology for Applications up to Ka Band," 2018 13th European Microwave Integrated Circuits Conference (EuMIC), Madrid, Spain, 2018, pp. 1–4, doi: 10.23919/EuMIC.2018.8539905.

[12] European Cooperation for Space Standardization, "ECSS-Q-ST-30-11C Rev.2: Derating – EEE components (23 June 2021).

[13] C. Ramella et al., "Thermal-aware GaN/Si MMIC design for space applications," 2019 IEEE International Conference on Microwaves, Antennas, Communications and Electronic Systems (COMCAS), Tel-Aviv, Israel, 2019,

[14] T. A. Milligan, "Modern Antenna Design," in Wiley-IEEE Press, 2nd ed., 2005

Francesco Manni received the M.S.Eng. degree in electronics engineering (summa cum laude) from the University of Roma Tor Vergata, Rome, Italy, in 2023, and is currently working toward the Ph.D. degree in space systems and technologies at the University of Rome Tor Vergata.

His current research interests include RF power amplifier theory, design and test, linearization techniques, and efficiency-improving techniques.

Paolo Colantonio (Fellow, IEEE) received the Laurea degree in electronics engineering and the Ph.D. degree in microelectronics and telecommunications from the University of Rome Tor Vergata, Rome, Italy, in 1994 and 2000, respectively. He is currently a full professor of microwave electronics with the University of Rome Tor Vergata. He has authored or co-authored more than 300 scientific articles. He has also authored the book titled *High Efficiency RF and Microwave Solid State Power Amplifiers* (Wiley, 2009), three book chapters, and four contributions to Wiley Encyclopedia on Microwave Electronics and holds one international patent. His research activities are mainly focused on the field of microwave and millimeter-wave electronics, and, in particular, on the design criteria for nonlinear microwave subsystems and high-efficiency power amplifiers. Dr. Colantonio has been the Chair for EuMIC 2022 and is currently an associate editor for the IEEE Microwave and Wireless Letters.

Rocco Giofrè (Senior Member, IEEE) received the Ph.D. degree in electronics from the University of Rome Tor Vergata, Rome, Italy, in 2008. He joined the Electronics Engineering Department, University of Rome Tor Vergata, in 2009, where he is currently an associate professor of electronics. He has authored more than 200 peer-reviewed articles, two book chapters, and two contributions for the Wiley Encyclopedia of Electrical and Electronics Engineering. His research activities belong to the microwave and millimeter-wave electronics area ranging from active device characterization to the design and test of linear and nonlinear circuits and systems. In this wide research area, he is mainly focused on the development of innovative power amplifier schemes and architectures with high efficiency and linearity for both ground and space communication systems, including their integration in multifunctional chips such as single-chip front-ends. He is involved in many research projects funded by international research agencies, such as the European Space Agency (ESA) and the Research Executive Agency (REA), European Commission, Brussels, Belgium. Dr. Giofrè is a member of the IEEE MTT-S Subcommittee 12 on power amplifiers, an associate editor for IEEE Access Journal, and a reviewer for the major journals and conferences of the field. He was a recipient of the 2005 Young Graduated Research Fellowship presented by the GAAS Association and the Best Paper Award at the EuMIC 2007.

Ernesto Limiti (Senior Member, IEEE) was a research and teaching assistant (1991–1997), an associate professor (1998–2001), and since then has been a full professor of electronics with the EE Department, University of Roma Tor Vergata, Rome, Italy. He has authored or co-authored over 500 publications on refereed international journals and presentations within proceedings of international conferences. His research activity is focused on three main lines, all of them belonging to the microwave and millimeter-wave electronics area. The first one is related to characterization and modeling for active and passive microwave and millimeter-wave devices. Regarding active devices, the research line is oriented to small-signal, noise, and large-signal modeling. For active devices, novel methodologies are developed for the noise characterization and the subsequent modeling, and equivalent-circuit modeling strategies are implemented both for small- and large-signal operating regimes for GaAs, GaN, SiC, Si, and InP MESFET/HEMT devices. The second line is related to design methodologies and characterization methods for low-noise devices and circuits. The main focus is on cryogenic amplifiers and devices. Finally, the third line is in the analysis and design methodologies for linear and nonlinear microwave circuits. Prof. Limiti is a referee for international journals of the microwave and millimeter-wave electronics sector. He is a member of the Steering Committee of international conferences and workshops. He has been the president of the Laurea and Laurea Magistrale degrees in electronic engineering of the University of Roma Tor Vergata, and is a member of the committee of the Ph.D. program in telecommunications and microelectronics, tutoring an average of four doctoral candidates per year. He is currently the Head of Electronic Engineering Department and the director of the Microwave Engineering Center for Space Applications (MECSA).

Patrick E. Longhi (Member, IEEE) received the M.S. degree in electronic engineering and the Ph.D. degree in microelectronics from the University of Roma Tor Vergata, Italy, in 2004 and 2009, respectively. From 2009 to 2018, he was with ELT – Elettronica Group, Rome, Italy, holding several managerial and technical positions in engineering, manufacturing, and sales. Since August 2018, he has been an assistant professor with the Electronic Engineering Department, the University of Roma Tor Vergata. He has authored more than 80 peer-reviewed articles and two book chapters. His main fields of research interest include analysis and synthesis methodologies for linear and noisy active networks at microwave and millimeter-wave frequencies, design of signal conditioning circuits, and design of multi-functional circuits for AESAs. He acts as Work Package Manager in research and innovation projects funded by the EU, ESA, or nationally (MIUR, ASI, Lazio Innova). He is a member of the IEEE MTT-S Technical Committee TC-11 on "Low Noise Techniques."

Steven Caicedo Mejillones received his Diploma in telematics engineering in 2014 and his master's degree in telecommunications in 2017, both from the Escuela Superior Politecnica del Litoral (ESPOL), ´ Guayaquil, Ecuador. He received the Ph.D. degree in information technology from the Politecnico di Milano, Milan, Italy, in 2023. From 2014 to 2018, he worked as a planning and optimization engineer for radio access networks (RAN) in different telecommunications companies in Ecuador, like the mobile operator Claro from America Movil Group. From 2018 to 2021, he was an early-stage researcher at SIAE MICROELETTRONICA in Milan, Italy, within the H2020 Marie-Curie ITN 5G STEP FWD program. Since 2021, he has been a microwave designer for space, backhaul, and O-RAN applications at SIAE. His research interests include synthesis and design techniques for microwave filters, filtering antennas, and phased array antennas.

Stefano Moscato (S'12) was born in Pavia, Italy, in 1988. He received the Ph.D. degree in electronics engineering from the University of Pavia, Pavia, Italy, in 2016. He was a visiting Ph.D. student at Georgia Tech, Atlanta, GA, USA, in early 2015. He became part of the R&D microwave group of SIAE MICROELETTRONICA in May 2017. His research activities have been focused on RF-to-mmWave passive components. Since September 2022, Dr. Moscato has been the coordinator of the 1337 R&D group devoted to the design and validation of mmWave passive components, antennas, and sub-systems. He is involved in innovation programs and founded researches for microwave backhauling, O-RAN equipment, and space-oriented assemblies. He was a recipient of an IEEE MTT-S Undergraduate/Pre-Graduate Scholarship in 2012. He is author of more than 50 papers on international journals and conferences. He has been the Chair of the IEEE Student Branch, University of Pavia, from 2013 to 2016.

Alessandro Fonte received the M.S. and Ph.D. degrees in electronic engineering from the University of Pisa, Italy, in 2005 and 2008, respectively. Since 2011, he has been with the SIAE MICROELETTRONICA S.p.A., Milan, Italy, where he is a senior microwave engineer and member of the technical staff. His current research interests include the design of micro- and millimeter-wave system-on-chip front-ends in silicon and III-V semiconductor technologies for point-to-point radio link applications. He has authored several papers in peer-reviewed international journals and conference proceedings and two book chapters. He is the SIAE contact person for European and ESA projects.

Appendix

Speaker name: Romain Hersent
Organization: III-V Lab
Country: France
Presentation title: InP on Si Technologies for Next-Generation Optical Communication High-speed Analog Front-Ends

SHORT ABSTRACT OF PRESENTATION

To achieve transmission speeds exceeding 1 Tb/s per channel at low cost/bit, it is essential to generate electrical multi-level pulse-amplitude modulated signals (PAM-4/8) at extremely high symbol-rates (>100 GBaud) [1]. Current digital systems heavily rely on silicon-based digital-to-analog converters (DACs), which analog bandwidth typically caps around 30–40 GHz [2]. Consequently, their signal-to-noise ratio (SNR) is often sacrificed to digitally enhance their bandwidth. This necessitates a DAC with a high effective number of bits at half the Nyquist frequency (>3–4 typically), leading to increased footprint and power consumption of the DAC IC. Along with the silicon CMOS data converters, silicon germanium (SiGe) linear amplifiers are often used to drive the optical modulators. However, these SiGe drivers suffer from a restricted bandwidth × output swing of less than 180 GHz.Vpp [3]-[4]. As a consequence, optical transmitter front-ends are either based on (i) sub-driven high-bandwidth modulators, which restrict the link reach and/or symbol-rate due to the subsequent optical SNR penalties, or (ii) low-Vpi modulators, compromising the overall bandwidth and/or footprint depending on which technology is used. Furthermore, power hungry digital signal processing (DSP) is frequently used to address various impairments, including bandwidth limitations, which significantly increases transceivers' power consumption, the link latency, and the analog signal peak-to-average power ratio (PAPR). Higher PAPR demands higher linearity in the analog front-end, thereby affecting its power consumption. Moreover, for next-generation optical systems operating beyond 200 GBaud, severe bandwidth degradations are caused by chip-to-chip interconnections and packaging [4], requiring extra DSP for compensation.

However, InP DHBT technologies have shown a high potential to bridge this performance gap, leveraging their very-high cutoff frequencies and large output power capabilities [5], [6]. Indeed, in [7], using InP DHBTs, the monolithic integration of an analog multiplexer (AMUX) and a linear modulator driver has demonstrated record gain-bandwidth product and output swing. This integrated circuit (IC) can aggregate two DACs bandwidth and double their sampling-rate to generate extremely high symbol-rate arbitrary signals, while directly driving the modulator. Moreover, leveraging the same technology, an electro-optical bandwidth surpassing 85 GHz has been achieved in [8], in employing an InP AMUX-driver and a thin film lithium niobate modulator. This assembly has shown 100-GBaud PAM-4 optical signals without any support of DSP. Additionally, InP DHBTs have enabled the realization of modulator driver ICs with performances exceeding 300 GHz.Vpp [9], [10]. Nonetheless, their integration onto silicon remains a challenge, to date, posing limitations on their widespread industrial adoption.

Within the Move2THz project, several paths are being explored to achieve seamless integration of InP technologies onto Si, within commercially viable frameworks. This is expected to substantially expand the opportunities for InP devices to serve as the backbone of future communication systems' analog front-ends.

During our presentation, we discuss the challenges related to the design of large-swing, high-efficiency, and extreme-high-symbol-rate analog electronics for next-generation optical communication systems offering more than 1.6 Tb/s, as well as for 6G mobile systems. Additionally, we illuminate the potential of InP-on-Si technologies to enhance signal integrity, as well as to improve thermal management and packaging.

KEYWORDS

Indium phosphide (InP), InP on silicon (InPoSi), analog multiplexer (AMUX), linear modulator driver, Tb/s optical communications, 6G, digital analog converter (DAC), digital signal processing (DSP)

ACKNOWLEDGMENTS

This work relates to the Chips Joint Undertaking (Chips JU) project Move2THz under grant agreement number 101139842. The Chips JU receives support from the European Union's Horizon Europe research and innovation program and the National Authorities.

REFERENCES

[1] Nokia.com "Nokia PSE-6s", Accessed: 11- Apr. 2023, [Online]. Available: https://www.nokia.com/networks/optical-networks/pse-6s/.

[2] T. Drenski and J. C. Rasmussen, "ADC & DAC—Technology trends and steps to overcome current limitations," in Proc. Opt. Fiber Commun. Conf. Expo. (OFC), 2018, pp. 1–3.

[3] H. Abdelrahman *et al.*, "A 6V swing 3.6% THD >40GHz driver with 4.5× bandwidth extension for a 272 Gb/s dual-polarization 16-QAM silicon photonic transmitter," 2019 IEEE ISSC proceedings, 2019, doi: 10.1109/ISSCC.2019.8662318

[4] J. Ozaki, Y. Ogiso, Y. Hashizume, H. Yamazaki, K. Nagashima and M. Ishikawa, "Coherent Driver Modulator With Flexible Printed Circuit RF Interface for 128-Gbaud Operations," in IEEE Photonics Technology Letters, vol. 34, no. 23, pp. 1289–1292, 1 Dec.1, 2022, doi: 10.1109/LPT.2022.3212678.

[5] A. M. Arabhavi *et al.*, "InP/GaAsSb Double Heterojunction Bipolar Transistor Emitter-Fin Technology

with fMAX = 1.2 THz," in IEEE Trans. on Electron Devices, vol. 69, no. 4, pp. 2122–2129, 2022. doi: 10.1109/TED.2021.3138379

[6] M. Urteaga *et al.*, "A 130 nm InP HBT integrated circuit technology for THz electronics," in 2016 IEEE International Electron Devices Meeting (IEDM), San Francisco, CA, USA, 2016, pp. 29.2.1–29.2.4, doi: 10.1109/IEDM.2016.7838503.

[7] R. Hersent *et al.*, "160-GSa/s-and-Beyond 108-GHz-Bandwidth Over-2-Vppd Output-Swing 0.5-μm InP DHBT 2:1 AMUX-Driver for Next-Generation Optical Communications," in IEEE Microwave and Wireless Components Letters, vol. 32, no. 6, pp. 752–755, June 2022, doi: 10.1109/LMWC.2022.3161706.

[8] R. Hersent *et al.*, "100 GBaud DSP-Free PAM-4 Optical Signal Generation Using an InP-DHBT AMUX-Driver and a Thin-Film Lithium Niobate Modulator Assembly," 2023 IEEE BiCMOS and Compound Semiconductor Integrated Circuits and Technology Symposium (BCICTS), Monterey, CA, USA, 2023, pp. 149–152, doi: 10.1109/BCICTS54660.2023.10310947.

[9] R. Hersent *et al.*, "InP DHBT Linear Modulator Driver With a 3-Vppd PAM-4 Output Swing at 90 GBaud: From Enhanced Transistor Modeling to Integrated Circuit Design," in IEEE Transactions on Microwave Theory and Techniques, doi: 10.1109/TMTT.2023.3305150.

[10] J. -Y. Dupuy et al., "A Large-Swing 112-Gb/s Selector-Driver Based on a Differential Distributed Amplifier in InP DHBT Technology," in IEEE Transactions on Microwave Theory and Techniques, vol. 61, no. 1, pp. 517–524, Jan. 2013, doi: 10.1109/TMTT.2012.2226057.

BIOGRAPHY

Romain Hersent (Member, IEEE) received the Engineering degree in electrical engineering from the ENSEA, Cergy, France, in 2016, and the Ph.D. degree in electrical engineering from L'Université de Cergy-Pontoise, Cergy, in 2020. He joined Nokia Bell Laboratories and III-V Lab, Nozay, France, in 2020, where he is currently a research engineer. His work focuses on the design and characterization of high-symbol-rate large-output-swing InP-DHBT integrated circuits for over 1 Tb/s/channel optical communications, and more specifically on analog multiplexers and linear drivers. Dr. Hersent is a member of the IEEE.

Speaker name: Kimon Vivien

Organization: United Monolithic Semiconductor

Country: France

Speaker picture:

Presentation title: **150 nm Gallium Nitride on Silicon Carbide Technology for High-power 5G New Radio Applications**

SHORT ABSTRACT OF PRESENTATION

In recent years, gallium nitride (GaN) transistors have become a key enabling technology for cellular communication networks, greatly contributing to the increase in energy efficiency and data rates of 5G new radio (5GNR) and future 6G. While LDMOS was the principal high power technology up until the 4G network, the very stringent constraints of modern telecommunications, as well as the opening to new frequency bands – mid-band (FR3) and millimeter waves (FR2) – are pushing operators to transition toward GaN.

As a company, United Monolithic Semiconductor (UMS) has been commercializing different GaN technologies on silicon carbide (SiC) substrates for the past decade, with gate lengths ranging from 500 down to 100 nm. With power densities reaching 4.5 W/mm at 30 GHz, associated with a drain efficiency above 50%, the 150 nm technological node GH15 is an optimal candidate for 5GNR, for all its frequency ranges.

Our presentation starts with discussing the main characteristics and options of GH15, along with some simulation and measurement data at key frequencies for 5GNR applications. The design and simulation results of a packaged 80 W Doherty power amplifier, offering more than 40% PAE at 9 dB back-off, is presented. This design was co-developed between UMS and STMicroelectronics within the SHIFT project. This design is currently under manufacturing – testing has been scheduled to start in September 2024. During the second part of our presentation, we focus on millimeter-wave FR2 power amplifiers. Some medium- and high-power UMS products compatible with 5G specifications are presented together with their challenges associated for 5GNR design, including linearity simulations and measurements. Then we conclude with presenting the latest technologies developed by UMS as well as a roadmap for the upcoming years, paving the way for future 6G networks.

KEYWORDS

5GNR, 6G, Doherty power amplifier, gallium nitride, linearity, silicon carbide, SHIFT project, UMS.

ACKNOWLEDGMENTS

This work is supported by the Chips Joint Undertaking (Chips JU) under grant agreement number 101096256 (project SHIFT). The Chips JU receives support from the European Union's Horizon Europe research and innovation program and the National Authorities.

BIOGRAPHY

Kimon Vivien (IEEE Member) received the Ph.D. degree from the University of Marnes-la-Vallée in 2020. In 2019 and 2020, he worked at the RF Power and Analog Laboratory, University of Colorado Boulder, USA, as an exchange Ph.D. student, under the supervision of Pr. Taylor Barton. Today, he holds the position of Senior RF and Microwave Engineer at United Monolithic Semiconductor, which he joined in 2020.

Speaker name: Joff Derluyn
Organization: Incize
Country: Belgium
Speaker picture:

Presentation title: **Post-process Substrate Porosification for RF Applications**

SHORT ABSTRACT OF PRESENTATION

The issue of the interaction of free charge carriers in the substrate with RF signals propagating through passive and active devices is well known [1]. This causes signal and efficiency losses, harmonic or intermodulation distortion, and cross-talk between devices. High-porosity, micro-porous silicon has been proposed as a potential solution to this problem [2], because the lateral dimensions of the remaining inter-pore silicon can be made small enough to induce a complete depletion, resulting in some cases in an effective resistivity above 100 kOhm.cm. Therefore, porous silicon is a promising technology for applications like low-noise amplifiers or RF switches. For higher power density devices (such as power amplifiers), the thermal conductivity of porous silicon may be too low to dissipate the power of the losses.

Typically, porous silicon is formed by an electro-chemical method prior to the fabrication of the CMOS circuitry. However, porous silicon is mechanically much more fragile than bulk silicon and it has a melting point of only 800 °C. Moreover, its sponge-like 3D structure easily traps contaminants and modifies the wetting properties, leading to poor cleanability. This makes the integration with various fab processes extremely challenging.

Instead, we have already proposed a novel approach to fabricate porous silicon in a post-CMOS-process approach [3]. This low temperature process only interacts with the backside of the wafers and is thus completely independent of the front-side layout and allows for local porosification. Excellent RF performance [4], including at high temperature up to 175 °C has been demonstrated [5]. The same technique was applied "post-epitaxy" to GaN-on-Si epitaxial wafers intended for RF applications [6], achieving RF losses under 0.1 dB/mm at 5 GHz and low second harmonic distortion H2 = −140 dBm at P_{out} = 15 dBm.

These results were obtained on highly p-doped Si substrates with resistivity below 20 mOhm.cm. Such low resistivity facilitates the porosification process, as the latter is essentially governed by hole exchange at the semiconductor-electrolyte interface: low resistivity substrates allow for a more uniform hole current and as a consequence a more uniform porous silicon layer. Unfortunately, the high boron doping in these substrates causes a concern for cross-contamination and therefore such substrates are not favored for introduction into RF-CMOS fabs.

In this work, we present a wafer-level, localized post-process porosification approach on semi-standard p-type silicon substrates with resistivity levels between 1 and 20 Ohm.cm. We will describe a number of modifications to the substrate layer stack and the electrode configuration that allow for the creation of homogenous porous silicon pockets on semi-standard substrates with an industry-relevant diameter. Small- and large-signal RF characterization data of effective resistivity, harmonic distortion, intermodulation distortion, and cross-talk will be presented.

REFERENCES

[1] M. Emam et al., "RF Semiconductor devices", Berlin, Germany, Springer (2013)

[2] G. Gautier, "RF Electrical Isolation with Porous Silicon", in "Handbook of Porous Silicon", Zug, Switzerland, Springer 2014, prt V, ch. 2

[3] G. Scheen et al., in IEEE MTT-S Int. Microw. Symp. Dig., Boston, MA, USA, Jun. 2019, pp. 1291–1294

[4] G. Scheen et al., "Post-process porous silicon for 5G applications", Solid State Electronics, vol 168 (2020), doi: 10.1016/j.sse.2019.107719

[5] M. Rack et al., "Small- and Large-signal Performance Up To 175°C of Low-Cost Porous Silicon Substrate for RF Applications", IEEE Trans. Electron Devices, vol. 65, no. 5, pp. 1887–1895, May 2018, doi: 10.1109/TED.2018.2818466.

[6] Scheen, Gilles et. al. GaN-on-Porous Silicon for RF Applications. 2023 53rd European Microwave Conference (EuMC) (Berlin, Germany, du 19/09/2023 au 21/09/2023). In: 2023 53rd European Microwave Conference (EuMC), , p. 842–845 (2023)

KEYWORDS

Harmonic distortion, intermodulation distortion, porous silicon, radio frequency, mmWave

BIOGRAPHY

Joff Derluyn (Member, IEEE) was born in Roeselare, Belgium, on March 13, 1974. He received the master's degree in electrical engineering from the University of Gent, Belgium, in 1998 and the Ph.D. degree in electrical engineering from the same institution in 2003 on the topic of epitaxy of dilute nitrides by MOCVD. He has worked at imec on the processing and characterization of GaN-based electronic devices. In 2010, he co-founded EpiGaN and served there as CTO until 2019 when EpiGaN was acquired by Soitec. At Soitec, he was a scientific director until 2022. He since joined Incize in Louvain-la-Neuve, Belgium, where he is a Fellow. He is the author of over 90 peer-reviewed papers or conference contributions and is co-inventor of 18 patent families. Dr. Derluyn's research interests focus on the interplay between material properties and semiconductor device physics.

Speaker name: Hassan Aboushady

Organization: Seamless Waves

Country: France

Speaker picture:

Presentation title: A Multi-standard RF Bandpass Sigma-Delta ADC

SHORT ABSTRACT OF PRESENTATION

Bandpass ΣΔ modulation is a very power-efficient technique to realize RF analog-to-digital converters (ADCs) [1]. To cover a wide frequency range, BP ΣΔ ADCs with tunable center frequencies have been proposed [2]. One important challenge in designing a tunable BP ΣΔ ADC relies in varying the modulator's feedback DAC coefficients to maintain the same signal transfer function (STF) and noise transfer function (NTF) for each center frequency, f0, in the complete tuning range, f0, of the ADC. The fact that the BP ΣΔ feedback DAC coefficients must be tuned for each center frequency has considerably complicated the modulator architecture. This resulted in a very large power consumption of 750 mW for a 1 GHz tuning range in [2], a rather low SNR of 40 dB with a bandwidth of 1 MHz for a 1.2 GHz tuning range in [3] or a very limited tuning range of 40 MHz in [4].

In a conventional BP ΣΔ architecture, the sampling frequency, fs, is fixed; so any variation in the center frequency, f0, leads to a variation in the normalized center frequency, f0/fs, and subsequently in the overall STF and NTF of the SD modulator. In this case, all the ΣΔ feedback DAC coefficients must be tuned to compensate for this f0/fs variation and maintain the same STF and NTF for all center frequencies. In this presentation, we present a tunable BP ΣΔ architecture, where the normalized center frequency, f0/fs, is fixed. In this case, any variation in the center frequency, f0, is tied to a variation in the sampling frequency, fs, to keep the normalized center frequency, f0/fs, equal to 1/4. Ideally in this case, all the ΣΔ feedback DAC coefficients can be fixed because f0/fs is fixed. Practically, varying the sampling frequency, fs = 1/Ts, changes the normalized loop delay, td/Ts, which leads to a modification of the modulator's NTF. To maintain the same NTF without tuning the feedback DAC coefficients, we propose to modify the loop delay, td, to compensate for any variation in the normalized loop delay td/Ts [5].

The proposed architecture is validated through the implementation of a second-order LC BP ΣΔ modulator in a 65 nm CMOS process. For an OSR of 64, the modulator achieves an Signal to Noise and Distortion Ratio (SNDR) of 37 dB over the entire tuning range, f0 = 1.5 GHz. This SNDR is achieved for a bandwidth BW = 47 MHz at f0 = 1.5 GHz and for a bandwidth BW = 93 MHz at f0 = 3.0 GHz. The fabricated chip consumes 13 mW and achieves the highest reported tuning range, f0, and the highest reported center frequency, f0, for a CMOS BP ΣΔ ADC.

In order to compare this circuit with the state of the art, we use not only the conventional ADC figure of merit, but we also use a figure of merit dedicated to RF circuits. In this case, the measured chip center frequency, noise figure, and non-linearity are also taken into account in the comparison.

KEYWORDS

Bandpass sigma-delta, CMOS, LC tanks, RF ADC, software-defined radio.

REFERENCES

[1] A. Sayed *et al.*, "A 10.8GS/s, 84MHz-BW RF Bandpass ΣΔ ADC with a 89dB-SFDR and a 62dB-SNDR for LTE/5G Receivers", IEEE VLSI Symposium 2024.

[2] H. Shibata *et al.*, "A DC-to-1 GHz tunable RF ΣΔ ADC achieving DR=74 dB and BW=150 MHz at f0 =450 MHz using 550 mW," IEEE JSSC, Dec. 2012.

[3] S. Gupta *et al.*, "A 0.8–2 GHz fully-integrated QPLL-timed direct-RF sampling bandpass ΣΔ ADC in 0.13 μm CMOS," IEEE JSSC, May 2012.

[4] H. Chae and M. P. Flynn, "A 69 dB SNDR, 25 MHz BW, 800 MS/s continuous-time bandpass ΣΔ modulator using a duty-cycle-controlled DAC for low power and reconfigurability," IEEE JSSC, Mar. 2016.

[5] H. Aboushady, T. Badran, and A. Sayed, "Tunable bandpass ΣΔ modulator," U.S. Patent 10 530 385, Jan. 2020.

BIOGRAPHY

Hassan Aboushady received the B.Sc. degree in electrical engineering from Cairo University, Egypt, in 1993, and the M.Sc. and Ph.D. degrees in electrical engineering and computer science from Sorbonne University, Paris, France, in 1996 and 2002, respectively. He has been an associate professor, currently on-leave, at Sorbonne University since 2003.

Hassan Aboushady is the co-founder and CEO of Seamless Waves, a spin-off from Sorbonne University, specialized in highly digitized RF transceivers. Dr. Aboushady has 30 years of experience in the field of RF/analog, ADCs, and DACs. He is the author and co-author of more than 90 publications and several patents in these areas. He was a visiting researcher at NXP, The Netherlands and at STMicroelectronics, France, in 1999 and 2001, respectively. Dr. Aboushady was a visiting professor at the French University in Egypt, the Federal University of Rio Grande do Norte, Brazil, Tecnologico de Monterrey, Mexico, and Ecole Polytechnique, France, in 2007, 2011, 2012, and 2013, respectively. Dr. Aboushady is a senior IEEE member, an IEEE-CAS distinguished lecturer, and a member of the IEEE Circuits and Systems for Communications Committee (CASCOM). He also served as an associate editor of the IEEE Transactions on Circuits & Systems II.

Speaker name: Cedric Dehos
Organization: CEA-Leti
Country: France
Speaker picture:

Presentation title: **D-band RF Architecture for Beyond 5G Wireless Networks: Specifications, Challenges, and Key Enabling Technologies**

SHORT ABSTRACT OF PRESENTATION

The D-band (110–170 GHz) could offer unprecedented data rates to mobile devices thanks to the use of huge non-contiguous frequency bands. Indeed up to 32 GHz band can be aggregated, allowing > 100 Gbps wireless data rate, for many short- or medium-range applications. Moreover, the most advanced BiCMOS technology (B55x) is pushing forward its maximum operating frequencies (ft/fmax), bridging the gap between pure SiGe and III-V technologies.

This mirage tends to vanish when trying to design transceivers and antennas at such frequencies, tackling big propagation (e.g. 96 dB at 10 m at 150 GHz) and implementation losses. If antenna array and beamforming scheme are suitable to provide antenna directivity, system gain and agility to transceiver at millimeter waves (mmWaves), the approach faces major issues when the frequency is so high. First the half-wavelength antenna pitch (~1 mm) imposes difficult constrains on the transceiver chip size, package assembly, antenna feeding routing, scalability, and heat dissipation. Next, the design of the RF phase shifters at D-band and over such wide bandwidth leads to awkward RF impairments, important losses, and huge power consumption.

In this context, this work focuses on the system definition and circuit specification on the "Future-G transceiver" demonstrator of the SHIFT project. After an analysis of the system requirement, we explore the architectural and technological options for the design of D-band transceivers and propose few key-enabling technologies for wireless access and backhauling, focusing on scalable and reconfigurable antenna schemes.

Channel aggregation and multi-tone frequency synthesis are first proposed to tackle the disseminated bands and bandwidth issues. Next, at the transmitter side, we describe a heterogeneous integration of a BiCMOS transmitter chip with III-V power amplifiers, and the beamforming antennas onto an RF interposer package to increase the radiation power. At the receiver side, a BiCMOS chip is optimized to reduce its noise figure and implementation losses, by the use of "in-local oscillators" phase shifters. Indeed, the transceiver local oscillators (Los) embed frequency multipliers (from a common frequency reference), raw delay-line phase shifters, and fine phase shifters based on an innovative injection locked oscillator (ILO) architecture. The ILO varactor is finely tuned to realize the desired phase shift according the Adler's theory. This approach reduces the losses in the RF path, provides versatility in the choice of the carrier frequency, and generates very low phase noise, allowing the transmission of higher order modulation schemes.

KEYWORDS

D-band, analog beamforming, LO phase shifters, injection locked oscillators

ACKNOWLEDGMENT

This work is supported by the Chips Joint Undertaking (Chips JU) under grant agreement number 101096256 (project SHIFT). The Chips JU receives support from the European Union's Horizon Europe research and innovation program and the National Authorities.

BIOGRAPHY

Cedric Dehos, born in 1980, joined CEA Leti, Grenoble, France, in 2003 after graduation from ESIEE Paris (Dipl.-Ing. and M.Sc. degrees) in telecommunication and signal processing. Since then, he has been involved in system-level design of complex RF and digital base band circuits including RF behavioral modeling, architecture design and specification, system simulation, and base band signal processing. His main topics of interest included ultra-wide band systems, digital compensation of RF impairments, and antenna beamforming. Since 2006, he has been involved in various mmWaves developments in CMOS or BICMOS technologies, including 24/79 GHz short range radars, 60GHz WiGig. In 2011, he led Leti mmw designs and began moving the developments towards 5G small cells, promoting the use of mmw bands for 5G, and towards short-range chip-to-chip communications and contactless connectors. More recently, he proposes architecture for channel aggregation in D-band, targeting >100 Gbps wireless communication. He has been involved in many collaborative projects funded by European Commission, as well as in many industrial bilateral projects with transfer of technology. He is also active in the promotion of mmw wireless technologies for particle physics with CERN.

Speaker name: Leonardo Gomes

Organization: Université Grenoble Alpes, ST Microelectronics

Country: France

Speaker picture:

Presentation title: **E-band and D-band VCOs: Distributed Tank Design Methodology, Bufferless Approach**

SHORT ABSTRACT OF PRESENTATION

Wireless communications have already moved into the millimeter wave (mmWave) range to satisfy the needs of an increasingly data-hungry society. To assure these speeds for the end-user, the backhaul, which carries the bulk of the data to be transmitted, needs to operate at even higher frequencies: to give the example of 5G, this happens at the E-band (71–76 GHz and 81–86 GHz).

Integrated circuit design at mmWaves relies heavily on the technology employed, both for active and passive devices. Nowadays, RF-oriented CMOS or BiCMOS technologies offer solutions to enable integrated circuit design at mmWaves, such as high-performance transistors and back-ends with thick upper metal layers for low loss passive devices. mmWave circuits are greatly impacted by interconnection parasitics, which oftentimes demands innovative solutions to enable a design or mitigate performance degradations. In the case of mmWave Voltage Controlled Oscillator (VCO) design accumulation, MOS varactors are the bottleneck of the tank performance, since their Q-factor is much lower than that of inductive resonators. However, varactor is the only passive device that is continuously variable and remains essential in VCO design. Moreover, MOS varactor losses are directly related to the channel area; so larger varactors, necessary for wide frequency tuning ranges (FTR), are more lossy than smaller varactors, thus evidencing an intrinsic design compromise between wide FTR and resonator quality factor.

A way to improve the tank performance consists in using distributed resonators based on slow-wave coplanar strip (S-CPS). Localized and periodically loaded with varactors topologies can be used. In order to accelerate the design, the electrical model and design methodology for such a resonator have to be developed, as well as how to size a MOSFET cross-coupled pair to realize loss compensation using a global optimization procedure on the loaded S-CPS resonator.

KEYWORDS

CMOS, distributed resonator, mmWaves, slow-wave coplanar stripline, standing-wave oscillator, SWO, VCO

ACKNOWLEDGMENT

This work is supported by the Chips Joint Undertaking (Chips JU) under grant agreement number 101096256 (project SHIFT). The Chips JU receives support from the European Union's Horizon Europe research and innovation program and the National Authorities.

BIOGRAPHY

Leonardo received the bachelor's degree in electrical engineering from the University of São Paulo, Brazil, in 2015, the M.Sc. degree in microelectronics also from the University of São Paulo, Brazil, in 2017, where he worked on designing of interposer antennas and passive devices at 60 GHz. In 2018, he enrolled into his Ph.D. in Cotutelle between the University of São Paulo and Université Grenoble-Alpes, France, where he worked on E-band, standing-wave oscillators, having also worked afterwards on frequency multipliers for D-band frequency generation and slow-wave interconnections for mmWaves on BiCMOS. He is currently working as an analog/RF design engineer at ST Microelectronics Crolles, France.

Speaker name: Valorge Olivier
Organization: CEA-Leti
Country: France
Speaker picture:

Presentation title: 2.5D, 3D Assembly Technologies for RF, mmW and Sub-THz Heterogeneous Systems

ABSTRACT

Heterogeneous integration is a must for future mmWave, sub-THz communication systems: there is fierce competition to develop components and systems with high data communication bandwidth. Combining III-V high speed/low-noise technologies with ultra-dense smart CMOS/BiCMOS technologies is a promising and exciting approach to investigate. Many technical challenges need to be addressed in different areas of physics: mechanical, electromagnetic, thermal. The solutions to these multi-physics problems must be cost-effective with a reasonable environmental impact in order to have a chance to be in pockets, cars, communication infrastructures, and others in a few years. CEA-Leti is investigating different 3D integration solutions that have been mastered in-house for tens of years, from the most mature one to the most advanced: chiplet copper pillar assembly and die to wafer direct hybrid bonding. This presentation focuses on different issues raised by mmW/sub-THz heterogeneous integrated systems and is illustrated by concrete examples.

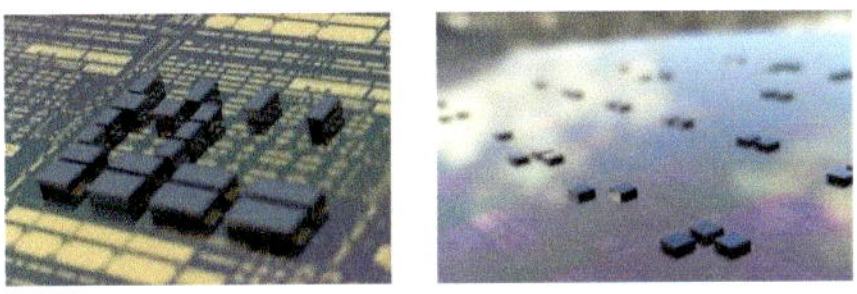

Fig. 1. 3D assembled demonstrators: (a) copper pillar assemblies; (b) direct hybrid bonding assemblies.

A passive silicon radio frequency (RF) interposer was designed and manufactured. It is the basis of many different passive and active RF-3D objects and demonstrators as shown in Figure 1. Top dies of different technologies can be reported on different bottom footprints, dedicated to RF measurements and characterizations, using different CEA-Leti 3D assembly techniques. Some passive structures, such as transmission lines in 2D and 3D contexts and their RF performances,

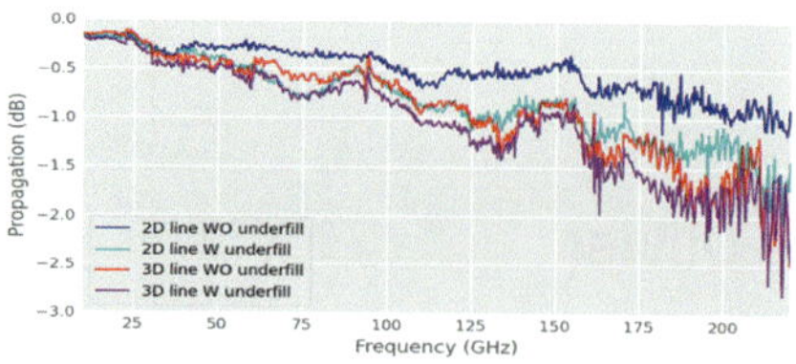

Fig. 2. Transmission line propagation in 2D and 3D contexts.

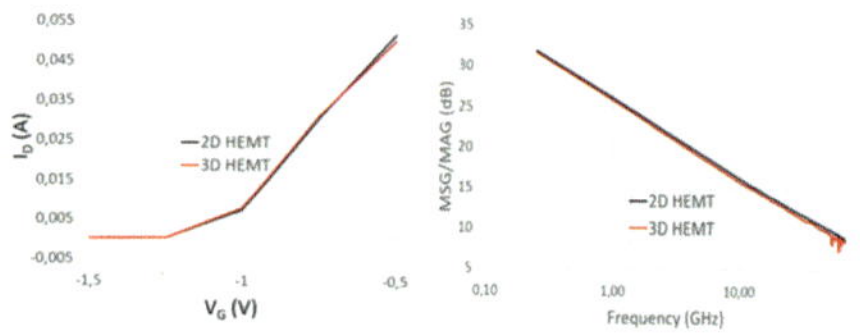

Fig. 3. DC and RF measurements of 2D vs. 3D GaN-High Electron Mobility Transistor (HEMT).

have been investigated (Figure 2). Several III-V devices were also assembled on the passive RF interposer: a GaN-RF HEMT, on which some electrical measurements are shown in Figure 3, and other indium phosphide (InP) Double Hetero-junction Bipolar Transistor (DHBT) using a similar 3D copper pillar assembly strategy will be presented. These different demonstrators allow evaluation of issues and optimizations to address to reach the electrical performances needed for future heterogeneous RF systems. The environmental impacts of the evaluated 3D assembly technologies are also studied and compared (Figure 4).

If the chiplet copper pillar assembly shows interesting results for short- and

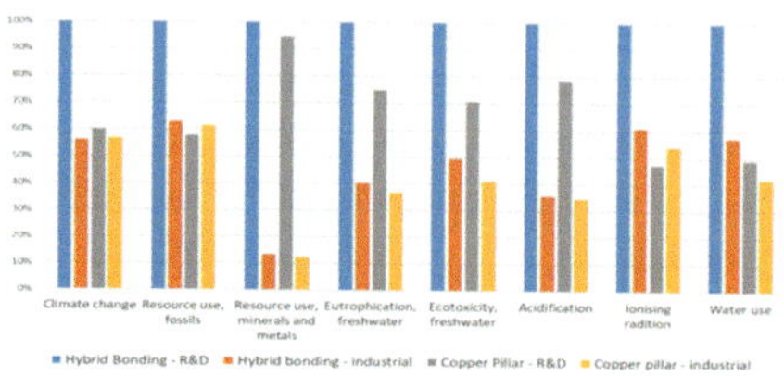

Fig. 4. Impact assessment comparison of hybrid bonding and CuPi.

medium-term 3D RF applications, optimizations and enhancements will allow the direct hybrid bonding process to open up new fields of possibilities for 3D RF systems in the medium and long terms.

KEYWORDS

Copper pillar, direct hybrid bonding, heterogeneous RF systems, impact assessment, passive RF interposer.

ACKNOWLEDGMENT

The presenter thanks the STMicroelectronics Crolles teams for fruitful discussions and financial support. This work was supported by the French Public Authorities ANR via Carnot Institute Funding and the IPCEI *Microelectronics and Connectivity* project within the frame of France 2030.

BIOGRAPHIES

Dr. Olivier Valorge joined CEA-Leti in September 2020. He received the Ph.D. degree in 2006 from INSA, Lyon, France, on substrate and power supply coupling in mixed-signal integrated circuits. This work was carried out in an industrial design environment: a mixed-signal design team at STMicroelectronics in Grenoble (France). Since 2006, his research has been focusing on possible applications of high technology devices in the semiconductor industry, in particular heterogeneous 3D integration for communication and lighting systems. He has worked as a researcher or research engineer in several academic, industrial, or start-up R&D units in France and Canada. His current research focuses on 3D heterogeneous integration for sub-THz applications and he coordinates CEA-Leti advanced internal projects

Dr. Christophe Dubarry received the M.Sc. degree in process engineering in 1996 from the University of Grenoble, France. He received the Ph.D. degree from University Joseph Fourier of Grenoble in 2001, with a thesis on the soft magnetic material CoFeCu for passive components. He joined CEA-Leti in 2001 in the framework of photonic devices development for laserdiscs. From 2008 to 2017, he worked in the new energy department for solar cells and microbatteries as a Process Integration Engineer and was involved in several European projects. Since 2017, he has been working at the Component on Silicon Department in CEA-LETI as a Device Integration Engineer and Project Leader. He is currently developing interconnection, in particular, bump and hybrid bonding for photonic and RF applications.

Dr. Hervé Boutry received the M.Sc. degree in materials sciences and microelectronics engineering in 1997 from the University of Lille, France. He received the Ph.D. degree from the Institute of Microelectronics and Nanoelectronics of Lille in 2002, with a thesis on the technological development of high electron mobility transistors based on Antimonide III-V based semiconductors for high-frequency low-noise amplifiers. After a few years at UCL-Louvain, Belgium, in the materials and physics laboratory, he joined CEA-Leti in 2004 in the framework of TFT transistors development for microdisplay cells. From 2008 to 2012, he worked in the Packaging and Interconnection Laboratory as a Process Integration Engineer and involved in several European projects. Since 2012, he has been working at the Component on Silicon Department in CEA-LETI as a Device Integration Engineer and Project Leader. He is currently developing an InP-based HBT technology integrated in Si-Fab platform.

Speaker name: Hady Yacoub
Organization: Ferdinand-Braun-Institut
Country: Germany
Speaker picture:

Presentation title: Heterointegration Approaches for InP-HBT Technologies for 5G Applications and Beyond

SHORT ABSTRACT OF PRESENTATION

The rapid developments in the field of monolithic microwave integrated circuits (MMICs) have been driven in the recent years by the increased demand for high-speed data rates for communications and sensing applications, in the optical, as well as in the wireless domain. Although Si-based technologies offer an extremely mature platform for the realization of complex systems, they lack key performance metrics such as high efficiency and output powers due to intrinsic material limitations.

III-V based semiconductors can deliver higher output powers with higher efficiencies; however, they lack the technology readiness levels of their silicon-based counterparts.

In the light of this conundrum, heterointegration emerged as a viable solution to maintain the advantages of both technology fields.

The proposed heterointegration approach relies on realizing the main functionality of the system using a more "mature" Si-based technology (that being CMOS or even BiCMOS), whilst replacing specific building blocks with III-V technology (in this case indium phosphide heterojunction bipolar transistors (InP-HBTs) MMIC process).

The system design through heterointegration adds one degree of complexity through the compatibility of the two technologies, but offers higher degrees of freedom for realizing significantly more efficient and compact RF systems.

The heterointegration technology that will be shown in this work is based upon a low-temperature bumping process, which uses indium as a bonding metal.

In contrast to the usual bumping process found in the backend-of-line processes, this process realizes the bumps in the frontend-of-line process of the InP MMIC process.

This gives a better edge in resolution and packing density, which in turn increases the heterointegration density.

The paper details results on the RF-interconnect properties of this technology and will show preliminary results of heterointegrated power amplifiers in InP on BiCMOS carriers.

Long-term stability as well as thermal cyclic tests have also been conducted to show the stability of this technology.

KEYWORDS

Bumps, chiplet, heterointegration, InP MMIC process, InP-HBT.

ACKNOWLEDGMENT

The authors would like to thank the financial support by the German BMBF within the "Forschungsfabrik Mikroelektronik Deutschland (FMD)" framework under ref. 16FMD02 and the project T-KOS (ref. 16KIS1405). The authors acknowledge the financial support by the Federal Ministry of Education and Research of Germany in the program of "Souverän. Digital. Vernetzt." joint project 6G-RIC, project identification numbers: 16KISK020K, 16KISK024, and 16KISK026. This work is related to Move2THz, a project supported by the Chips Joint Undertaking (Chips JU) under grant agreement number 101139842. The Chips JU receives support from the European Union's Horizon Europe research and innovation program and the National Authorities.

BIOGRAPHY

Hady Yacoub was born in 1987 in Giza, Egypt. He received the bachelor's degree in electrical engineering from the German University in Cairo in 2009. He received the M.Sc. degree in microelectronics and communication from the University of Ulm, Germany, in 2011. He later joined RWTH Aachen University, where he acquired his doctoral degree in 2017. He joined Ferdinand-Braun-Institut as a post-doctoral fellow in 2018.

Hady Yacoub currently leads InP devices lab since 2019. His research interests are fabrication and characterization of compound semiconductor devices for high-frequency and energy applications. He authored and co-authored more than 30 papers on various topics in III-V semiconductors.

Speaker name: Siddhartha Sinha
Organization: imec
Country: Belgium
Speaker picture:

Presentation title: **RF-Heterointegration at Wafer-level and Panel-level for mmWave Applications**

SHORT ABSTRACT OF PRESENTATION

Above 100 GHz packaging for millimeter-wave (mmWave) radars and phased arrays presents four fundamental packaging challenges:

- Reduction of antenna array pitch (e.g. 1 mm at 150 GHz), which requires the dense placement of antenna elements and electronics
- Increase in the size and types of chips needed: CMOS (beamforming) needs to be complemented with III-V (GaN, InP, GaAs) for final high-power low-noise stage
- Increase in metal and dielectric losses with frequency
- Thermal management as higher DC losses are generated in smaller areas leading to higher thermal densities

The packaging activities in imec's advanced-RF department aim to address these challenges and consist of two tracks.

The first track is a wafer-level packaging RF-interposer technology developed on 300 mm silicon substrates. This technology uses standard resistivity silicon substrates to reduce costs while achieving excellent mmWave and mixed-signal performance as the silicon substrate is shielded from the RF signals by using a ground plane. The main features of this technology are as follows:

- First, we incorporate a standard Cu damascene multi-layer backend process for digital interconnects for providing multiple parallel digital links in the low GHz range (1–3 GHz) but with high density (1um trace/space).
- Second, on top of the Cu damascene, we spin-coat two low RF-loss thick layers of polymer and semi-additive thick Cu Redistribution Layer (RDL)

layers. This provides an impedance matched mmWave link, which can provide broadband mmWave performance (DC – 200 GHz) at moderate densities (5 um trace/space, 30 um pitch).

- Third, high aspect ratio Through Silicon Via (TSV) (10:1), wafer thinning down to 50–100 um and TSV transitions through the silicon substrate to transfer mmWave signals to the backside of the interposer for 3D stacking and mounting.
- Fourth, we have also incorporated high performance and small footprint passives (MIMCAP up to 50 fF/um^2, high Q inductors, and TaN resistors) in the Cu damascene backend for supply decoupling at mmWave frequencies for GaN/InP/GaAs chiplets to be mounted on the interposer.
- Fifth, we have also demonstrated flip-chip interconnects for both CMOS (Cu backend) and III-V (GaN/InP/GaAs with Au backend) chiplets on our interposer with 40 um micro-bump pitch with broadband RF performance up to 110 GHz (to be extended further in frequency). Demonstrators for phased arrays and radars using this technology are being developed at imec-Florida, USA under a US National Science Foundation Engines grant.

The second track on panel-level packaging focuses on two activities:

- First, die embedding of mmWave chipsets in an organic core package and development of antenna-in-package technology. Die-embedding offers two benefits over flip-chip interconnect: (1) the connections to the antenna are shorter as the mmWave signal needs to traverse only the top half of the build-up and (2) the bottom half of the build-up can be used for thermal management.
- The second activity involves the development of air-filled substrate integrated waveguide (AFSIW) technology for mmWave applications where we have demonstrated an AFSIW with loss of 0.07 dB/mm at 140 GHz with solid sidewalls and broadband launcher with insertion loss of 1.1 dB.

Both activities are being continued further with partners of the advanced-RF program.

KEYWORDS

RF, heterointegration, CMOS, III-V, antenna, phased array, InP, GaN, GaAs, interposer, chiplet, TSV, flip-chip, die embedding, air-filled substrate integrated waveguide (AFSIW).

BIOGRAPHY

Siddhartha Sinha has been a part of the advanced-RF department of imec, Leuven, Belgium, since 2015. He is responsible for electromagnetic modeling, III-V/CMOS heterointegration, mmWave antennas, packaging and system technology co-optimization (STCO) of mmWave systems. Between 2010 and 2015, he was a scientist with the Ferdinand-Braun-Institut (FBH), Berlin, Germany, working on mmWave interconnects and equivalent circuit modeling for FBH's InP transistor technology. Between 2004 and 2006, he was a scientist at Defence R&D Organisation (DRDO), Bangalore, India, working on travelling wave tubes. He holds a bachelor's degree from Visweswaraya Technological University, India and a master's degree from Technical University Munich, Germany.

Speaker name: Frederic Gianesello
Organization: STMicroelectronics
Country: France
Speaker picture:

Presentation title: **SiGe BiCMOS & III-V Technologies Heterogeneous Integration Challenges**

SHORT ABSTRACT OF PRESENTATION

To address consumer needs for more ubiquitous mobility, integration has been driving the wireless business to achieve the appropriate cost and form factors. This led to the progressive replacement of historical III-V MMIC technology (such as GaAs) by Si-based ones (RF CMOS, RF-SOI, and BiCMOS). Moving to a more aggressive CMOS technology (in terms of gate length) enabled to either reduce the die size or embed more features for the same size in a system-on-chip. This approach led to current highly integrated 5G RF transceivers, which are now integrated in 12 nm FinFET technologies.

But aggressive gate length scaling is not the only advantage of Si technologies, and derivative Si technologies also play a key role. It is especially true for the RF-SOI technology that has replaced the GaAs technology to achieve RF switches integrated in most RF front-end modules. However, Si-based technologies are still used in a support role in RF front-end, which remains served by an assortment of technologies such as GaAs HBT, acoustic filters, integrated passive devices, surface mount devices, and Si-based technologies (all those technologies being integrated using a system-in-package approach).

Moving to 6G future wireless standard, next-generation RF Front End Module (FEM) will require a breakthrough in terms of performance and integration (while operating on wider band higher in frequency). To do so, cost-effective SiGe BiCMOS technology can once again take advantage of heterogeneous integration with III-V technologies to achieve the best performance/cost tradeoff. In this perspective, compound semiconductors such as GaN High Electron Mobility Transistor (HEMT) and InP Heterojunction bipolar transistor (HBT)/HEMTs are attracting a lot of attention. GaN HEMTs may address future challenges to integrate efficient power amplifiers and low-loss RF switch at millimeter-wave (mmW). Additionally, InP HBTs and HEMTs exhibit excellent linearity and noise performance up to mmW.

However, GaN and InP technologies have a relatively slow transition from laboratory to qualified process and associated manufacturing cost is way higher than Si-based technologies one. Consequently, the main strategy consists today in limiting the GaN and InP content to small chiplets containing only transistors, while the other features (for example, passive devices) are achieved using lower cost technologies. Dedicating the III-V wafer to transistors enables more efficient area usage, and, thus, yields more than 10,000 chiplets even on a 4” wafer.

This approach requires to disaggregate the circuit into an active compound semiconductor chiplet and a low-cost interposer. This paper will review different integration strategies that can be considered to support proposed heterogeneous integration schemes leveraging existing Si-based or advanced packaging technologies. It will allow us to identify the current technical challenges and limiting factors to be addressed in the coming years to enable proposed disaggregated heterogeneous integration to become a reality.

KEYWORDS

6G mobile communication, heterogeneous integration, SiGe BiCMOS, GaN HEMT, InP HBT, InP HEMT, system in package.

ACKNOWLEDGMENT

This work is related to the SHIFT and Move2THz projects supported by the Chips Joint Undertaking (Chips JU) under grant agreement number 101096256 and number 101139842. The Chips JU receives support from the European Union’s Horizon Europe research and innovation program and the National Authorities.

BIOGRAPHY

Fred Gianesello received the B.S. and M.S. degrees in electronics engineering from Institut National Polytechnique de Grenoble, Grenoble, France, in 2003, and the Ph.D. degree in electrical engineering from the Joseph Fourier University, Grenoble, France, in 2006.

Dr. Gianesello has authored and co-authored more than 200 refereed journals and conference technical articles. He has served the TPC of IEEE International SOI Conference, IEEE International Solid-State Circuits Conference, IEEE International Electron Device Meeting Conference, and IEEE EuMIC Conference and is currently serving the TPC of IEEE RFIC.

Dr. Gianesello has been working for ST on the development of silicon-based RF and mmW technologies for the past 20 years targeting RF and mmW transceivers and front-end module products. He is currently working for STMicroelectronics in Crolles, France, as technical expert providing support to customer and performing business prospection for RF, mmW, and optical products.

Speaker name: Tanja Braun
Organization: Fraunhofer IZM
Country: Germany
Speaker picture:

Presentation title: **Advanced Packaging Solutions for mmWave Applications**

SHORT ABSTRACT OF PRESENTATION

With the end of Moore's law, advanced packaging and heterogeneous system integration is gaining more importance. At the same time in the last years, the increasing demand for ever higher data rates in wireless communication has exposed the limits of current sub-6 GHz communication systems and pushed the technology toward the exploration of the millimeter-wave (mmWave) frequency spectrum. In particular, three frequency bands have been licensed at 26, 28, and 39 GHz for the new generation of mobile communication (5G), each exhibiting as much as 3 GHz of operational bandwidth. However, systems operating at mmWave must cope with much higher free space losses than in the sub-6 GHz frequency range. Moreover, the physical size of interconnects (chip-to-chip, chip-to-antenna) becomes comparable to the operating wavelength. Therefore, new packaging technologies are required, which allow implementing large antenna arrays for massive MIMO and short interconnects to minimize package losses. Radar systems working at 77 GHz for, e.g., autonomous driving have also entered the market. All these new applications also require new packaging solutions allowing higher manufacturing volume and lower cost.

Advanced packaging solutions suitable for RF and mmWave applications are, e.g., flip chip and interposer solutions, fan-out wafer level packaging (FOWLP), or chiplet approaches. This also addresses topics such as antenna-in-package solutions, shielding on package level, passive component integration, or heterogeneous integration of III/V semiconductors with Si or SiGe in one package.

From the above-mentioned technologies, FOWLP is currently one of the strongest trends in microelectronics packaging. FOWLP has a high potential in significant package miniaturization concerning package volume but also in thickness reduction. The main advantages of FOWLP are the substrate-less package, low thermal resistance, and high RF performance due to shorter interconnects. Especially inductance of FOWLP is much lower compared to FC-BGA (Flip Chip Ball Grid Array) packages. In addition, the redistribution layer can also integrate embedded passives (R, L, and C) as well as antenna structures using a multi-layer structure. It can be used for multi-chip packages for system in package, package stacking, and heterogeneous integration. Hence, technology is well suited for RF applications.

FOWLP is applied for volume production for mobile and wireless applications (mainly wireless baseband) and has meanwhile been introduced in automotive applications. Infineon started with a 77 GHz Radar IC Chip Set being the first 77 GHz solution based on FOWLP eWLB (embedded Wafer Level Ball Grid Array) instead of a bare die. As the technology offers a lot of opportunities for RF packaging with shortest interconnect lengths and the possibility of integrating passive components and even antenna structures, more applications will enter the market, especially with the advent of 5G and 6G applications.

KEYWORDS

Advanced packaging, antenna-in-package, fan-out wafer level packaging.

BIOGRAPHY

Tanja Braun studied mechanical engineering at Technical University of Berlin with a focus on polymers and micro systems and joined Fraunhofer IZM in 1999. In 2013, she received her Dr. degree from the Technical University of Berlin. Tanja Braun is head of the department of System Integration and Interconnection Technologies. Her recent research is focused on fan-out wafer and panel level packaging technologies. In 2021, she received the Exceptional Technical Achievement Award from IEEE Electronics Packaging Society (EPS) and the IMAPS Sidney J. Stein Award for her work in the field of fan-out wafer and panel level packaging. Tanja Braun is an active member of IEEE. She is a member of the IEEE EPS Board of Governor (BOG) and is the IEEE EPS VP of Conferences.

Speaker name: Mikko Varonen

Organization: VTT Technical Research Centre of Finland Ltd.

Country: Finland

Speaker picture:

Presentation title: **Modular 3D mmW and THz Packaging Concepts and Technologies**

SHORT ABSTRACT OF PRESENTATION

This talk will start with a short overview of different monolithic microwave integrated circuit (MMIC) technologies for millimeter-wave (mmW) and THz operation followed by the design of MMIC phased array circuits. We will discuss packaging techniques for efficient phased array scaling and demonstrate a 3D modular packaging concept for heterogeneous integration of different MMIC technologies. The 3D modular packaging concept is based on an active transmit array antenna and is scalable in size due to quasi-optical feeding and stackable printed circuit board design. We also show the progress of VTT's wafer-level packaging platform integrating antennas, filters, and active MMIC chips on a single silicon carrier. The technology is based on micromachined waveguides in silicon, Au-Au wafer bonding, and flip-chip technologies. The process includes only standard fabrication steps that can be automated and allows integration of individual MMICs of different technologies into the same Si module. The process is scalable and cost efficient for future sub-THz and THz communication and sensing applications.

KEYWORDS

Heterogeneous integration, integration, micromachining, phased arrays, terahertz (THz).

ACKNOWLEDGMENT

This work was supported by the Business Finland through Terafront (decision number 8398/31/2022) and RF Sampo projects and by the Academy of Finland through

MilliRAD (decision number 314542), TeraCom (decision number 319681), and MIDERI (decision number 310234) projects.

BIOGRAPHY

Mikko Varonen received the M.Sc., Lic.Sc., and D.Sc. (with distinction) degrees in electrical engineering from the Aalto University (formerly, the Helsinki University of Technology), Espoo, Finland, in 2002, 2005, and 2010, respectively.

In 2011, he was a NASA Postdoctoral Program Fellow with the Jet Propulsion Laboratory (JPL), California Institute of Technology (Caltech), Pasadena, CA, USA. In 2012, he was a post-doctoral researcher, and he was the Academy of Finland Post-Doctoral Researcher with the Department of Micro- and Nanosciences, Aalto University, from 2013 to 2016. During his Post-Doctoral Fellowship, he was a visiting scientist with the JPL, Electrical Engineering Department, Caltech, and the Fraunhofer Institute of Applied Solid-State Physics, Freiburg, Germany. He is currently a Principal Scientist with VTT Technical Research Centre of Finland, Espoo, where he held Academy of Finland Research Fellow post during 2017 to 2022.

His current research interests include the development of RF and mmW-integrated circuits using both silicon and compound semiconductor technologies for applications ranging from astrophysics and earth remote sensing to mmW communications and quantum computing.

Speaker name: Abdel Hadi Hobballah

Organization: S2P-Smart Plastic Products

Country: France

Speaker picture:

Presentation title: **LDS and AMP Processes for RF Antenna in Package (AiP) Applications in the E- and D-Bands**

SHORT ABSTRACT OF PRESENTATION

The SHIFT project develops sustainable technologies for future telecom applications by seeding an innovative European ecosystem. This part of this ecosystem is dedicated to establishing new manufacturing chains of intelligent microelectronics packaging building on innovative active mold packaging (AMP) technology.

In the SHIFT project, S2P-Smart Plastic Products is positioned on the manufacturing of the advanced packaging of components using the AMP technology, for antenna-in package (AiP) and antenna-in-module (AiM) solutions. The use of AMP technology has several advantages, including: (i) integration of antennas and interconnections in small volume, (ii) innovative radio frequency (RF) concepts, and (iii) carbon footprint reduction using recyclable packaging materials, and reduction of precious metals.

The AMP process is based on three main steps:

- the molding of thermoplastics or thermo-set epoxy mold compound;
- the laser activation to structure the artwork directly from the computer onto the plastic component;
- the chemical electroless plating to create the track.

The KMC9220 resin from Shin-Etsu was chosen due to its preliminary adoption by the packaging ecosystem. The material has been supplied as pellets and delivered to STMicroelectronics, Grenoble, France, which performed preliminary transfer molding trials. Using molded samples, we made first activation and metallization trials with a fiber 1064 nm infrared pulsed laser (3D-Microline 160i), to build the design rules for electronics and antennas designers using this equipment. On the other hand, we also tested a UV laser at 355 nm (Picoline 3000 Ci), which enables tracks and interspaces of 25 µm and lower roughness to address high-frequency applications.

In parallel with the mentioned AMP tests, the electrical material characterization was achieved by the IEMN Laboratory, Lille, France, by providing them some LDS resin samples. The obtained results show that the dielectric constant DK and the dissipation factor DF are about 3.8–3.9 and 0.005–0.01, respectively, in the frequency range of 100–600 GHz.

The material characterization will be validated by designing, manufacturing, and measuring some RF functions such as 50 Ω microstrip line and coplanar waveguide, before proceeding to the targeted AiP demonstrators in the project.

KEYWORDS

Antenna-in-package, epoxy resins.

ACKNOWLEDGMENT

SHIFT project receives funding from Chips Joint Undertaking (Chips JU) under grant agreement number 101096256. The Chips JU receives support from the European Union's Horizon Europe research and innovation program and the National Authorities.

BIOGRAPHY

Abdel Hadi Hobballah received the M.Sc. degree in applied physics and physical engineering, and the Ph.D. degree in antenna and RF circuit design from the University of Limoges, France, in 2019 and 2022, respectively. He is currently a project manager with S2P-Smart Plastic Products, France, where he applies his RF skills in the manufacturing of antennas using the laser direct structuring (LDS) process of the 3D-MID molded interconnect device technology.

Speaker name: Akanksha Bhutani
Organization: Karlsruhe Institute of Technology
Country: Germany
Speaker picture:

Presentation title: **Sub-THz Antenna and Package Integration for Miniaturized Surface-Mount Device Modules**

SHORT ABSTRACT OF PRESENTATION

The availability of a wide usable bandwidth above 100 GHz makes the use of these frequencies highly appealing for 6G wireless communication and high-resolution millimeter-wave (mmWave) radars. However, despite significant advancements in semiconductor technologies enabling monolithic microwave integrated circuits (MMICs) operating above 100 GHz, their utilization has largely remained limited to laboratory experiments and low-volume applications. This limitation is primarily due to challenges related to limited transmission power and higher path loss, which have been widely acknowledged. To fully unlock the potential of these frequencies, there is an urgent need to develop broadband, high-efficiency antennas, and low-loss interconnects between antennas and MMICs. Moreover, factors such as heat dissipation, cost, and size further complicate the implementation of high-performance mmWave and terahertz (THz) systems. One critical aspect that demands attention is the dilemma between off-chip and on-chip solutions for systems operating above 100 GHz, each having its own advantages and challenges. Therefore, this talk focuses on two key research areas: (1) the development of off-chip antenna and package concepts in the WR6 band (110–170 GHz) using mature and novel packaging technologies, and (2) the advancement of high-gain and high-efficiency on-chip antennas in the WR3 (220–325 GHz) and WR2.2 (325–500 GHz) bands, utilizing innovative techniques in combination with different semiconductor technologies. In summary, this talk sheds light on the advancements made in mmWave and THz antenna and package technologies while highlighting the remaining challenges that must be overcome to accelerate the adoption of higher frequency bands in future radar and wireless communication systems. By addressing these obstacles, we can pave the way for realizing the full potential offered by frequencies beyond 100 GHz.

KEYWORDS

Antenna, 6G, IC packaging, millimeter-wave, sub-THz, radar, wireless communication.

ACKNOWLEDGMENT

This work is financially supported by the Federal Ministry of Education and Research (BMBF), Germany under the project "ESSENCE-6GM" (grant number: 16KISK173) and "Open6GHub" (grant number: 16KISK010).

BIOGRAPHY

Akanksha Bhutani received the M.Sc. and Ph.D. degrees in electrical engineering and information technology from the Karlsruhe Institute of Technology (KIT), Germany, in 2012 and 2019, respectively. From 2012 to 2019, she served as a research assistant with the Institute of Radio Frequency Engineering and Electronics (IHE), KIT. Since 2021, she has been leading the Antennas and Packaging research group at IHE, KIT. Her primary research revolves around THz antennas and packaging for radar and wireless communication systems. She received the "Carl Freundenberg Prize" and the "Suedwestmetall Advancement Award" for her dissertation in 2019 and 2020, respectively. Additionally, she received the IEEE Microwave Magazine Best Paper Award in 2017 and the European Microwave Week (EuMW) Best Paper Awards in 2019 and 2022. In 2023, she served as the Operations Officer at EuMW 2023, held in Berlin. Later that year, she was honored with the prestigious International IHP "Wolfgang Mehr" Fellowship Award by the Leibniz-Institut für innovative Mikroelektronik (IHP). She has authored and co-authored around 60 research papers published in peer-reviewed conference proceedings and journals.

Index

For Product Safety Concerns and Information please contact our EU representative GPSR@taylorandfrancis.com Taylor & Francis Verlag GmbH, Kaufingerstraße 24, 80331 München, Germany